A Companion to Peripheral Nervous System Examination

Antonino Uncini • Roberto Eleopra
Paolo Girlanda • Fiore Manganelli
Lucio Santoro

A Companion to Peripheral Nervous System Examination

Antonino Uncini
Department Neuroscience, Imaging and
Clinical Science
University of Chieti-Pescara
Chieti, Italy

Paolo Girlanda
Department Clinical and Experimental
Medicine
University of Messina
Messina, Italy

Lucio Santoro
Department of Neurosciences Reproductive
Sciences and Odontostomatology
University of Naples Federico II
Napoli, Italy

Roberto Eleopra
Department Clinical Neurosciences
Istituto Neurologico Carlo Besta
Milano, Italy

Fiore Manganelli
Department of Neurosciences Reproductive
Sciences and Odontostomatology
University of Naples Federico II
Napoli, Italy

ISBN 978-3-031-63627-1   ISBN 978-3-031-63628-8   (eBook)
https://doi.org/10.1007/978-3-031-63628-8

© Società Italiana di Neurofisiologia Clinica - SINC 2025

This work is subject to copyright. All rights are solely and exclusively licensed by the Publisher, whether the whole or part of the material is concerned, specifically the rights of translation, reprinting, reuse of illustrations, recitation, broadcasting, reproduction on microfilms or in any other physical way, and transmission or information storage and retrieval, electronic adaptation, computer software, or by similar or dissimilar methodology now known or hereafter developed.

The use of general descriptive names, registered names, trademarks, service marks, etc. in this publication does not imply, even in the absence of a specific statement, that such names are exempt from the relevant protective laws and regulations and therefore free for general use.

The publisher, the authors and the editors are safe to assume that the advice and information in this book are believed to be true and accurate at the date of publication. Neither the publisher nor the authors or the editors give a warranty, expressed or implied, with respect to the material contained herein or for any errors or omissions that may have been made. The publisher remains neutral with regard to jurisdictional claims in published maps and institutional affiliations.

This Springer imprint is published by the registered company Springer Nature Switzerland AG
The registered company address is: Gewerbestrasse 11, 6330 Cham, Switzerland

If disposing of this product, please recycle the paper.

Preface

The Italian Society of Clinical Neurophysiology (Società Italiana di Neurofisiologia Clinica, SINC) is pleased to present: *A Companion to Peripheral Nervous System Examination*. This manual is configured as a practical, quick, bedside reference with the intent to help the trainee in the clinical and electrophysiological examination of the peripheral nervous system.

The manual contains three parts.

The Part I is a brief introduction to the clinical examination of the peripheral nervous system with reference figures and tables that are helpful to localize the lesion site.

In the Part II, after a concise explanation of some basic principles on nerve conduction studies, are illustrated the most common techniques to perform motor, sensory, and mixed nerve conduction studies as well as some special studies. The description follows the same schematic structure with the indication of the nerve fibers tested, their route and the placement of recording and stimulating electrodes. Control values for each technique as well exemplificative recordings are reported. Comments on technical pitfalls and clinical correlations are also provided.

In the Part III, after an introduction to needle electromyography, it is described, and shown by photographs, how to test the strength and activate the muscles that are more frequently examined. The needle insertion point is shown and described using, when appropriate, anatomical landmarks. Moreover, by means of anatomical cross sections, oriented according to the viewpoint of the examiner, the relationships between the muscle to be examined and adjacent muscles and structures are demonstrated remarking how to avoid common pitfalls. This section contains also, for a rapid consultation, diagrams of plexuses and peripheral nerves with the muscles which they supply.

The authors hope that this handbook will be of some practical value for those approaching the sometimes difficult, though often rewarding, electrodiagnosis of the peripheral nervous system.

Antonino Uncini

On behalf of the Italian Society of Clinical Neurophysiology (SINC)

Contents

Part I   Principles of Peripheral Nervous System Testing

1   Clinical Examination 3
   1.1   Symptoms 3
   1.2   Muscle Testing 3
   1.3   Sensory Examination 7
   1.4   Tendon Reflexes 14
   References 15

Part II   A Guide to Routine Nerve Conduction Studies

2   Basic Principles of Nerve Conduction Studies 19
   2.1   Potential Recording 19
   2.2   Nerve Stimulation 20
   2.3   Motor Conduction Studies 20
   2.4   Sensory Conduction Studies 22
   2.5   Temperature and Nerve Conduction 24
   2.6   Measurement of Distance 25
   2.7   Reference Values 25

3   Cranial Nerve Studies 27
   3.1   Facial Nerve Study 27
   3.2   Facial Branches Study 29
      3.2.1   Frontal Branch 29
      3.2.2   Zygomatic Branch 30
      3.2.3   Mandibular Branch 30
      3.2.4   Facial Branches Stimulation in Studying
              the Lateral Spread of Excitation 31
   3.3   Blink Reflex Study 33
   3.4   Spinal Accessory Nerve Study 36
   References 38

4   Cervical, Brachial, and Upper Limb Nerve Studies 39
   4.1   Phrenic Nerve Study 39
   4.2   Long Thoracic Nerve Study 42

4.3 Suprascapular Nerve Study 45
4.4 Axillary Nerve Study 47
4.5 Musculocutaneous Nerve Study 49
4.6 Lateral Cutaneous Nerve of the Forearm Study 51
4.7 Radial Nerve Motor Study 53
4.8 Posterior Cutaneous Nerve of the Forearm Study 55
4.9 Radial Nerve Sensory Study 57
4.10 Medial Cutaneous Nerve of the Forearm Study 59
4.11 Ulnar Nerve Motor Study 61
4.12 Deep Ulnar Motor Branch Study 64
4.13 Ulnar Nerve Sensory Study to Digit 5 66
4.14 Ulnar Dorsal Cutaneous Nerve Study 69
4.15 Anterior Interosseous Nerve Study 71
4.16 Median Nerve Motor Study 73
4.17 Median Nerve Sensory Study to Digit 2 or 3 75
    4.17.1 Antidromic Technique 76
    4.17.2 Orthodromic Technique 76
4.18 Median to Ulnar Comparative Studies 78
    4.18.1 Median Versus Ulnar Digit 4 Sensory Study 79
    4.18.2 Median Versus Ulnar Mixed Nerve Study
           from Palmar Stimulation 84
    4.18.3 Median 2nd Lumbrical Versus Ulnar Interossei
           Distal Motor Latencies Study 86
4.19 Median Motor and Sensory Segmental Studies 88
    4.19.1 Motor Segmental Study 88
    4.19.2 Sensory Segmental Study 90
4.20 Yield of Comparative and Segmental Studies in CTS 92
4.21 An Electrophysiological Classification of CTS 93
References 94

5 **Lower Limb Nerve Studies** 99
5.1 Femoral Nerve Study 99
5.2 Lateral Femoral Cutaneous Nerve Study 101
5.3 Saphenous Nerve Study 103
5.4 Peroneal Nerve Study Recording from the Extensor
    Digitorum Brevis Muscle 105
5.5 Peroneal Nerve Study Recording from the Tibialis
    Anterior Muscle 108
5.6 Superficial Peroneal Sensory Study 110
5.7 Tibial Nerve Study 112
5.8 Medial and Lateral Plantar Nerve Motor Studies 114
5.9 Medial and Lateral Plantar Nerve Sensory Studies 117
5.10 Sural Nerve Study 119
References 121

6	Special Studies	123
	6.1 F Wave Studies	123
	6.1.1 Median and Ulnar Nerves	123
	6.1.2 Peroneal and Tibial Nerves	124
	6.2 Soleus H Reflex Study	127
	6.3 Bulbocavernosus Reflex Study	130
	References	131

Part III A Guide to Needle Electromyography

7	Needle Electromyography (EMG)	135
	7.1 Basic Principles	135
	7.2 EMG Evaluation During Electrode Insertion and at Rest	136
	7.3 EMG Evaluation at Minimal Voluntary Effort	137
	7.4 EMG Evaluation at Maximal Voluntary Effort	138
	7.5 Steps in Needle EMG	138
	7.6 Anatomy for Electromyography	139
	Selected References	139

8	Muscles Innervated by the Cranial Nerves	141
	8.1 Trigeminal Nerve, Masseter	141
	8.2 Facial Nerve	142
	8.2.1 Frontalis	143
	8.2.2 Orbicularis Oculi	144
	8.2.3 Mentalis	145
	8.3 Spinal Accessory Nerve	147
	8.3.1 Sternocleidomastoid	147
	8.3.2 Upper Trapezius	148
	8.4 Hypoglossal Nerve, Musculi Linguae	149

9	Muscles Innervated by Nerves of the Brachial Plexus	151
	9.1 Long Thoracic Nerve, Serratus Anterior	152
	9.2 Dorsal Scapular Nerve, Rhomboideus Major	153
	9.3 Medial and Lateral Pectoral Nerves, Pectoralis Major	155
	9.4 Suprascapular Nerve	156
	9.4.1 Supraspinatus	156
	9.4.2 Infraspinatus	158
	9.5 Thoracodorsal Nerve, Latissimus Dorsi	160

10	Muscles Innervated by the Musculocutaneous, Axillary, and Radial Nerves	163
	10.1 Musculocutaneous Nerve, Biceps Brachii	163
	10.2 Axillary Nerve	166
	10.2.1 Teres Minor	167
	10.2.2 Deltoid	169

    10.3   Radial Nerve 170
           10.3.1   Triceps Brachii-Caput Laterale 170
           10.3.2   Brachioradialis.. 172
           10.3.3   Extensor Carpi Radialis (ECR) Longus............... 173
           10.3.4   Extensor Digitorum Communis (EDC) 174
           10.3.5   Extensor Indicis Proprius (EIP)..................... 176

11   Muscles Innervated by the Median Nerve 179
    11.1   Pronator Teres 180
    11.2   Flexor Carpi Radialis (FCR) 181
    11.3   Flexor Digitorum Sublimis (FDS)............................. 183
    11.4   Flexor Digitorum Profundus (FDP)........................... 184
    11.5   Flexor Pollicis Longus (FPL)................................. 186
    11.6   Pronator Quadratus 187
    11.7   Abductor Pollicis Brevis (APB)............................... 189
    11.8   Opponens Pollicis 190

12   Muscles Innervated by the Ulnar Nerve 193
    12.1   Flexor Carpi Ulnaris (FCU).................................... 194
    12.2   Abductor Digiti Minimi (ADM) 195
    12.3   Interosseous Dorsalis I.. 196
    12.4   Flexor Pollicis Brevis (FPB) 198

13   Muscles Innervated by the Femoral and Obturator Nerves......... 201
    13.1   Femoral Nerve... 203
           13.1.1   Iliacus... 203
           13.1.2   Rectus Femoris....................................... 204
           13.1.3   Vastus Lateralis 206
    13.2   Obturator Nerve, Adductor Magnus.......................... 207

**14   Muscles Innervated by the Superior
       and Inferior Gluteal Nerves**................................... 209
    14.1   Superior Gluteal Nerve 210
           14.1.1   Tensor Fasciae Latae (TFL)............................ 210
           14.1.2   Gluteus Medius 211
    14.2   Inferior Gluteal Nerve, Gluteus Maximus..................... 213

15   Muscles Innervated by the Sciatic and Tibial Nerves 215
    15.1   Biceps Femoris-Caput Longum (BF-CL) 216
    15.2   Semimembranosus... 217
    15.3   Gastrocnemius-Caput Mediale (GCM) 219
    15.4   Soleus.. 220
    15.5   Tibialis Posterior 222
    15.6   Abductor Hallucis Brevis (AHB).............................. 223
    15.7   Abductor Digiti Quinti Pedis (ADQP)......................... 225

16	**Muscles Innervated by the Sciatic and Peroneal Nerves**	**227**
	16.1 Biceps Femoris-Caput Brevis (BF-CB)	228
	16.2 Peroneus Longus	229
	16.3 Tibialis Anterior	231
	16.4 Extensor Digitorum Longus (EDL)	232
	16.5 Extensor Hallucis Longus (EHL)	233
	16.6 Extensor Digitorum Brevis (EDB)	235
17	**Paraspinal Muscles**	**237**
	17.1 Cervical	237
	17.2 Thoracic	238
	17.3 Lumbar	239

Abbreviations

A	Active electrode
CMAP	Compound muscle action potential
CTS	Carpal tunnel syndrome
CV	Conduction velocity
DML	Distal motor latency
EMG	Needle electromyography
ENMG	Electroneuromyography
G	Ground electrode
LLN	Lower limit of normal
m/s	Meter/second
mA	Milliampere
ms	Millisecond
MUP	Motor unit potential
mV	Millivolt
N	Nerve
PNS	Peripheral nervous system
R	Reference electrode
SNAP	Sensory nerve action potential
ULN	Upper limit of normal
μV	Microvolt

Part I

Principles of Peripheral Nervous System Testing

Clinical Examination

The main goal of the neuromuscular examination is to localize the disorder to a specific component of the peripheral nervous system (PNS). Always keep in mind that nerve conduction studies and needle electromyography (EMG) are an extension of the clinical examination; a brief history and a focused examination should always be performed. The duration, type, and distribution of symptoms, along with the physical examination, help to formulate the differential diagnosis, which is at the basis of the electroneuromyography (ENMG) plan.

1.1 Symptoms

The main symptoms in patients with neuromuscular disorders are weakness, sensory disturbances, and pain in varying combinations. Patients with sensory abnormalities most commonly report a reduction or loss of sensation in skin areas described as numbness and medically termed hypesthesia, or sensations of pins and needles, tingling, prickling, and burning, commonly termed paresthesias. Getting a precise localization of the abnormal sensation is an important step toward a diagnostic localization. However, not all patients can delineate accurately the affected area. For example, patients with carpal tunnel syndrome (CTS) often describe numbness of all fingers of the hand, and only if they are asked to think about it and be more precise do they describe a more restricted distribution. Motor symptoms also vary. Patients may complain of muscle twitching, cramps, clumsiness in a limb, or weakness.

1.2 Muscle Testing

Muscle testing is an essential part of the neuromuscular examination. A muscle may act as a prime mover, synergist, fixator, or antagonist. Testing individual muscles in relation to the movement of a single joint, rather than muscle groups, is important for diagnosis in peripheral nervous system (PNS) disorders. When a one-joint and a multi-joint muscle act together in a movement, the action of the one-joint muscle can be differentiated by placing the multi-joint muscle at a mechanical disadvantage. For example, in testing the hip extension by gluteus maximus, the action of hamstrings is decreased by flexing the knee so that the hamstrings are in their most shortened position. Regarding synergistic muscles, for example, to isolate the action of the pronator quadratus from the pronator teres, the elbow can be flexed to induce the maximal shortening of the pronator teres, making its action less effective. In the third section of the manual, for brevity, only one method of testing for each muscle is shown, and we recommend applying the test as illustrated to eliminate possible pitfalls. The strength of muscles should be always quantified. The simplest and most widely used grading scale for muscle weakness is that of the British Medical Research Council (Table 1.1).

Grades 4−, 4, and 4+ may be used to indicate movement against slight, moderate, and strong resistance, respectively.

Examining the strength of individual muscles helps to localize the lesion. This can be done by referring to diagrams of roots, plexuses, and peripheral nerve supply of muscles. Figures 1.1 and 1.2 show the most common patterns of segmental spinal root innervation of muscles (myotomes) in upper and lower limbs. However, these charts should be used with some caution because of individual variation of muscle innervation by roots. Moreover, regarding the possibility of attributing any clinical or EMG pattern to a specific root level, it should be reminded that muscles are seldom innervated by only one spinal root.

Table 1.1 Medical Research Council (MRC) scale for the evaluation of muscle strength

0 = no muscle contraction
1 = visible contraction not resulting in movement
2 = active movement with gravity eliminated
3 = active movement against gravity
4 = active movement against moderate resistance
5 = normal strength

1.2 Muscle Testing

Nerve/Muscles	C5	C6	C7	C8	T1
Dorsal scapular nerve					
Rhomboid major/minor	R				
Suprascapular nerve					
Supraspinatus	R	Y			
Infraspinatus	R	Y			
Axillary nerve					
Deltoid	R	R			
Musculocutaneous nerve					
Biceps Brachii	R	R			
Median nerve					
Pronator teres		R	R		
Flexor carpi radialis		R	R		
Flexor pollicis longus			Y	R	Y
Abductor pollicis brevis				R	R
Ulnar nerve					
Flexor carpi ulnaris			Y	R	Y
Abductor digiti minimi				R	R
First dorsal interosseous				R	R
Radial nerve					
Triceps		Y	R	Y	
Brachioradialis	R	R			
Extensor carpi radialis longus/brevis		R	R		
Extensor digitorum communis			R	Y	
Extensor indicis proprius			Y	R	

Fig. 1.1 This chart shows the most useful muscles to examine clinically and electromyographically in radiculopathies of the upper limb arranged according to nerve and root innervation. Red boxes indicate the predominant roots innervating the muscle, yellow boxes indicate the contribution by other roots

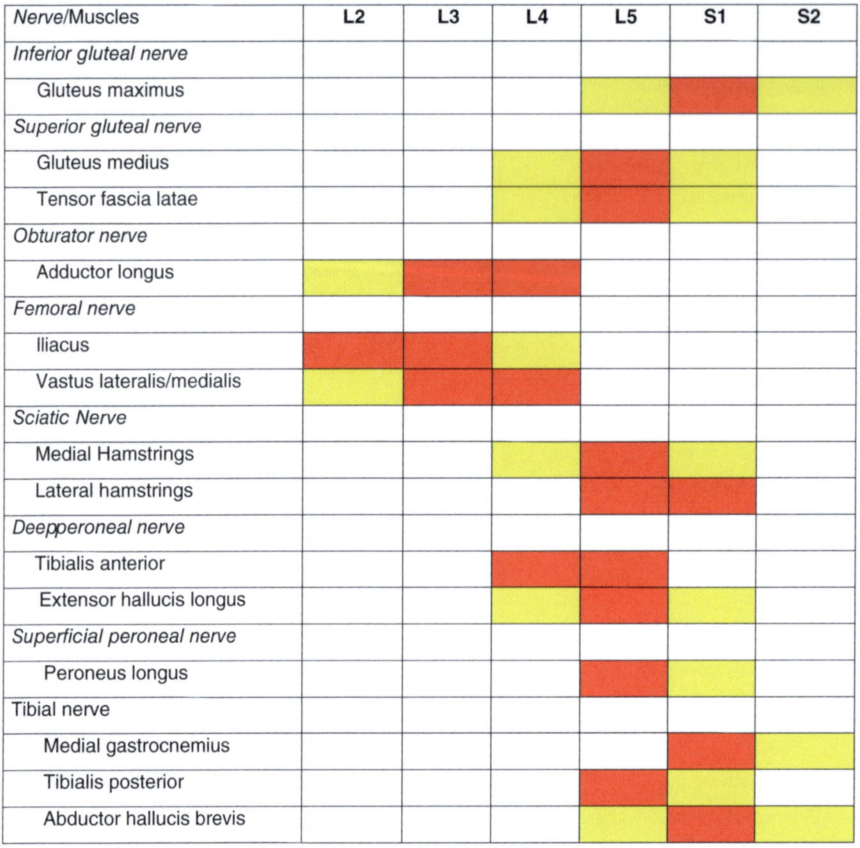

Nerve/Muscles	L2	L3	L4	L5	S1	S2
Inferior gluteal nerve						
Gluteus maximus				yellow	red	yellow
Superior gluteal nerve						
Gluteus medius			yellow	red	yellow	
Tensor fascia latae			yellow	red	yellow	
Obturator nerve						
Adductor longus	yellow	red	red			
Femoral nerve						
Iliacus	red	red	yellow			
Vastus lateralis/medialis	yellow	red	red			
Sciatic Nerve						
Medial Hamstrings				red	red	yellow
Lateral hamstrings				red	red	
Deep peroneal nerve						
Tibialis anterior			red	red		
Extensor hallucis longus			yellow	red	yellow	
Superficial peroneal nerve						
Peroneus longus				red	yellow	
Tibial nerve						
Medial gastrocnemius					red	yellow
Tibialis posterior				red	yellow	
Abductor hallucis brevis				yellow	red	yellow

Fig. 1.2 This chart shows the most useful muscles to examine clinically and electromyographically in radiculopathies of the lower limb arranged according to nerve and root innervation. Red boxes indicate the predominant roots innervating the muscles, yellow boxes indicate the contribution by other roots

Diagrams of plexuses and peripheral nerves with the sequence of the branches that innervate the muscles are reported in the third part of the manual. There are several pitfalls in motor examinations. Weakness may be unreliable because the patient may be confused, inattentive, uncooperative, malingering, or having a functional neurological disorder. Coexisting pain may interfere with strength testing, masking weakness; on the other hand, if weakness is also complained of, the relief of pain may demonstrate that muscle strength is normal.

1.3 Sensory Examination

Sensory testing takes time and needs good cooperation from the patient. The purpose is to identify sensory abnormalities and delimit precisely the site, the extent, and the sensory modality affected. The sensory examination begins by asking the patient to delineate the area of sensory abnormality. Then, with the patient's eyes closed, light touch can be tested by something soft such as cotton wool or a piece of paper, but the examiner's finger is frequently sufficient by checking from the insensitive toward the sensitive area. Superficial pain is tested by a pin moving from the analgesic area outward. The size of the insensitive pinprick area is usually inferior to the one of light touch. The timing of stimulation should be irregular so that the patient does not know when to expect the next touch or pinprick. It is useful to compare light touch and pinprick sensation in abnormal and normal areas on the same side (i.e., median versus ulnar innervated digits of the same hand) or with the same area of the unaffected side. Occasionally testing for two-point discrimination by a caliper, or simply by the two points of an unfolded paper clip, can reveal subtle sensory abnormalities not detectable by light touch or pinprick. The normal two-point discrimination varies in different body parts, and in the fingertips, it should be no more than 5 mm. Testing vibration sense with a 64 or 124 Hz tuning fork placed on various bony prominences is also useful in evaluating peripheral neuropathies. The distribution of sensory loss is usually partial except in the most severe nerve lesion. For example, in CTS, paresthesias and sensory loss are usually confined only to the fingertips rather than to the whole cutaneous distribution of the median nerve, possibly because the longest nerve fibers are more susceptible to compression. In other CTS patients, sensory abnormalities may be confined to one digit. This is possibly due to predominant damage of specific fascicles within the nerve. The conventional representation of skin areas supplied by spinal nerve roots (dermatomes) is shown in Fig. 1.3. It should be reminded that the skin area supplied by any nerve root varies considerably from one individual to another and dermatomes overlap considerably. Therefore, it is unusual in an isolated radiculopathy to develop a severe sensory disturbance in the whole dermatome.

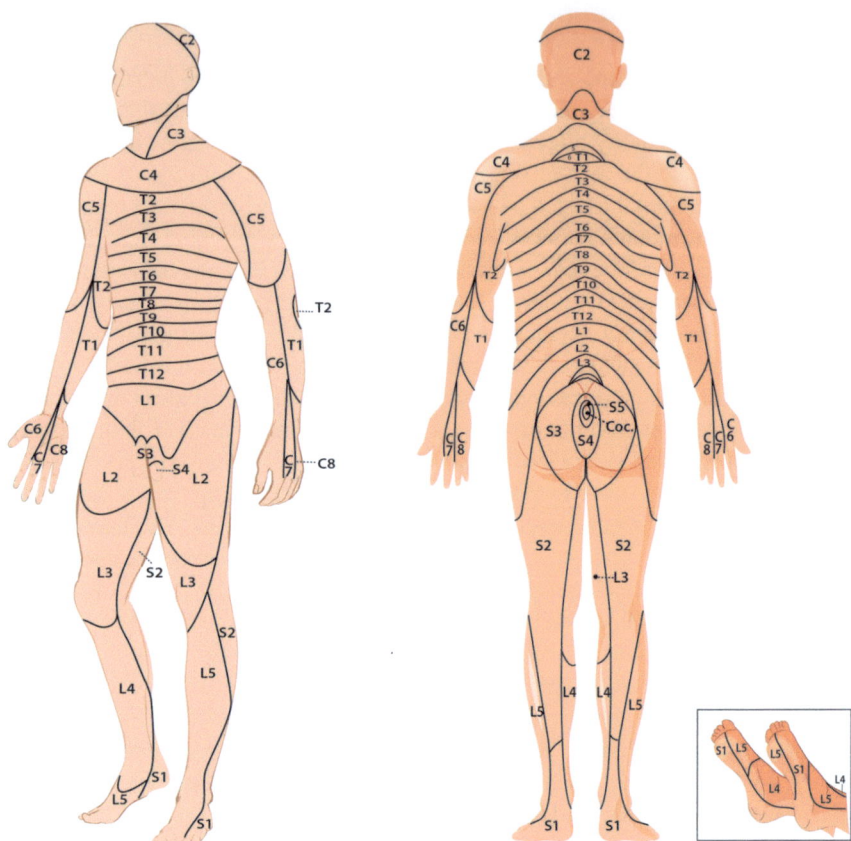

Fig. 1.3 The dermatomes of the body

Dense numbness and sensory loss usually are more indicative of a peripheral nerve lesion. The cutaneous distribution of the nerves of the body is shown in Fig. 1.4.

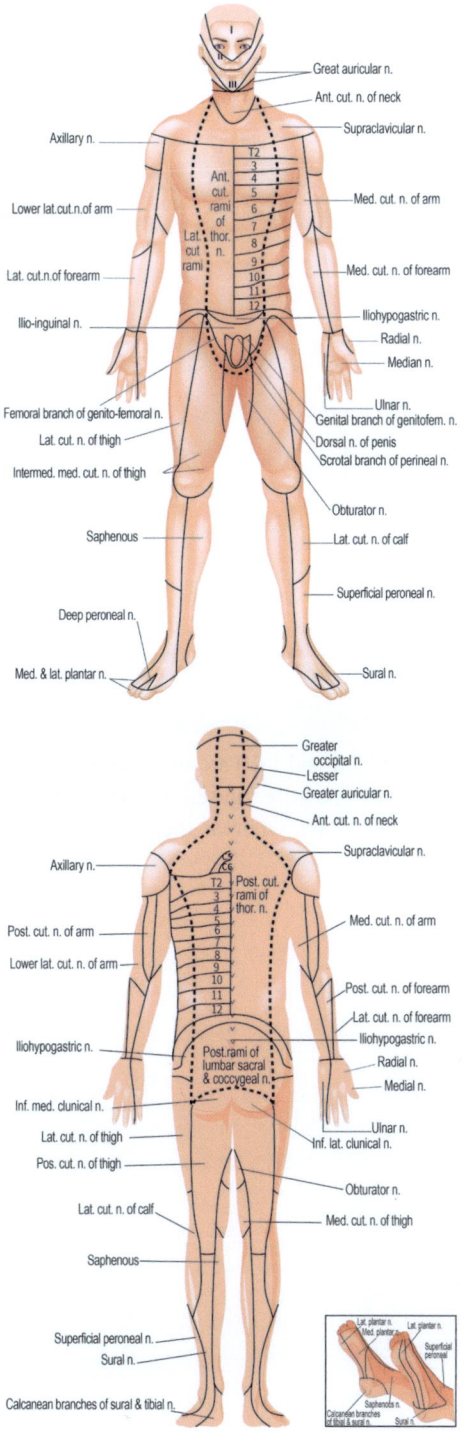

Fig. 1.4 The cutaneous distribution of the nerves of the body

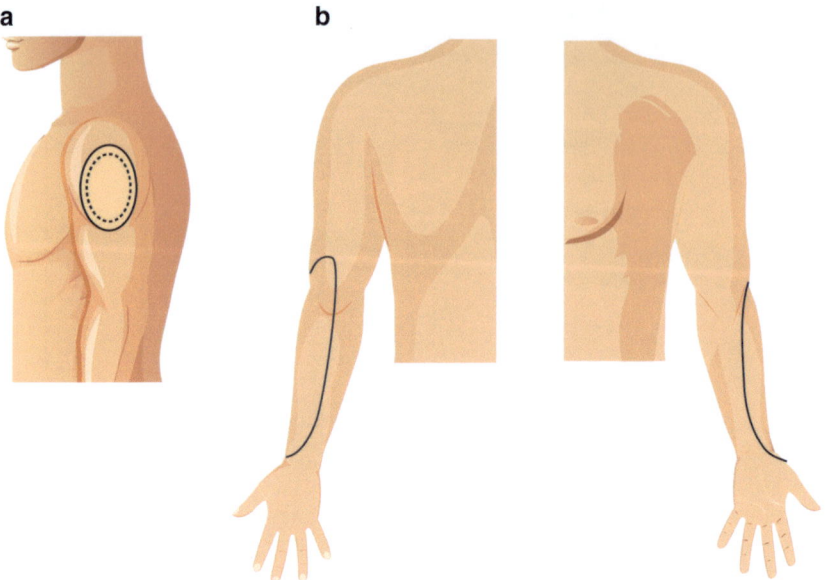

Fig. 1.5 The approximate area within which sensory changes can be found in lesions of (**a**) the axillary nerve and (**b**) the musculocutaneous nerve (lateral cutaneous nerve of the forearm), posterior and anterior view

In Figs. 1.5, 1.6, 1.7, 1.8, 1.9 and 1.10 are reported the approximate area of sensory loss in the nerves most frequently injured. In these figures, the continuous line individuates the loss of light touch, whereas the dotted line identifies the loss pinprick sensation.

1.3 Sensory Examination

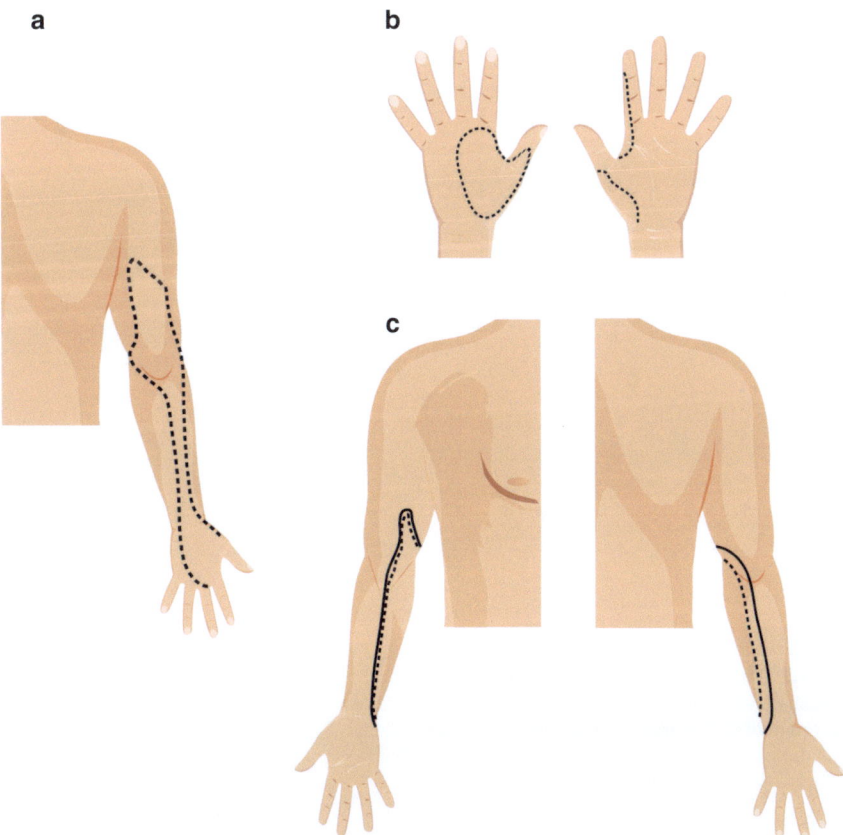

Fig. 1.6 The approximate area within which sensory changes can be found in lesions of (**a**) the radial nerve above the origin of the posterior cutaneous nerve of the arm and forearm; (**b**) the radial nerve below the origin of the posterior cutaneous nerve of the forearm, dorsal and volar view; and (**c**) the medial cutaneous nerve of the forearm, anterior and posterior view

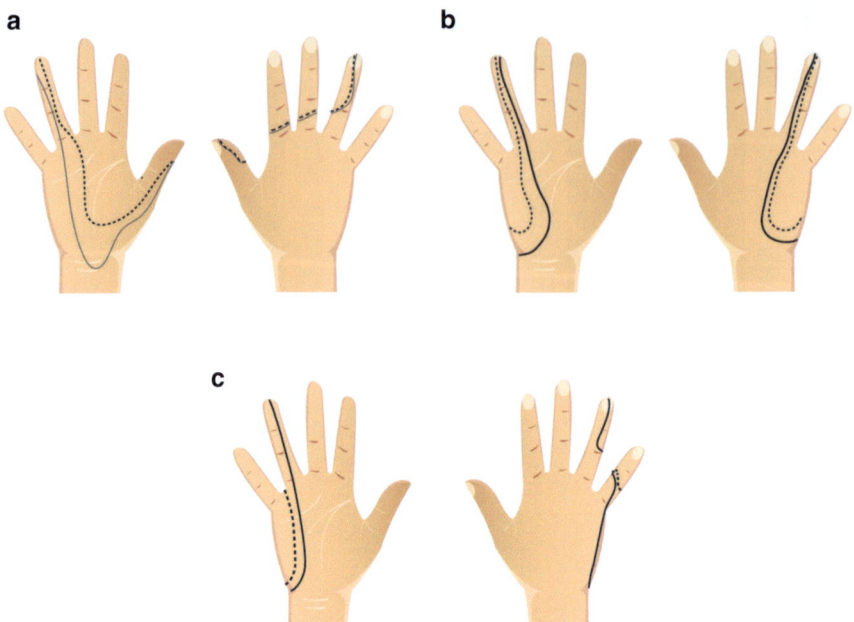

Fig. 1.7 The approximate area within which sensory changes may be found in lesions of (**a**) median nerve, volar and dorsal view; (**b**) ulnar nerve, volar and dorsal view; (**c**) ulnar nerve below the origin of dorsal branch, volar and dorsal view

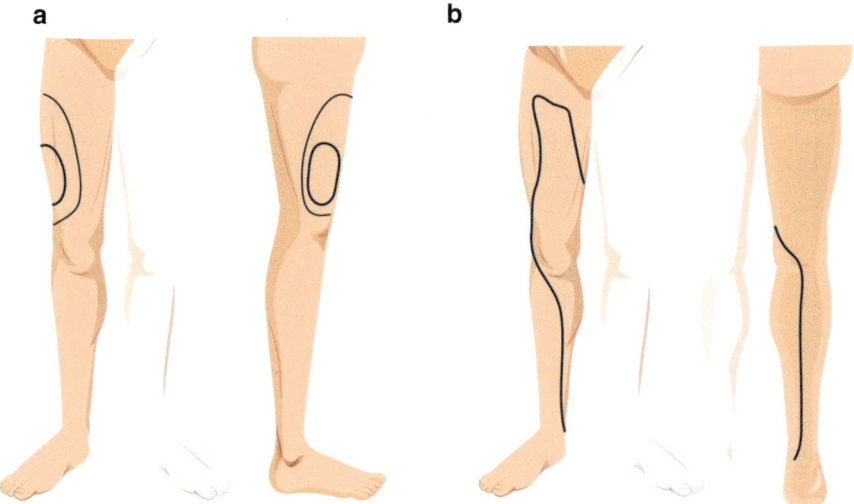

Fig. 1.8 The approximate area within which sensory changes may be found in lesions of (**a**) lateral femoral cutaneous nerve, anteromedial and lateral view; (**b**) femoral nerve (intermediate and medial cutaneous nerve of the thigh and saphenous nerve), anteromedial and posterior view

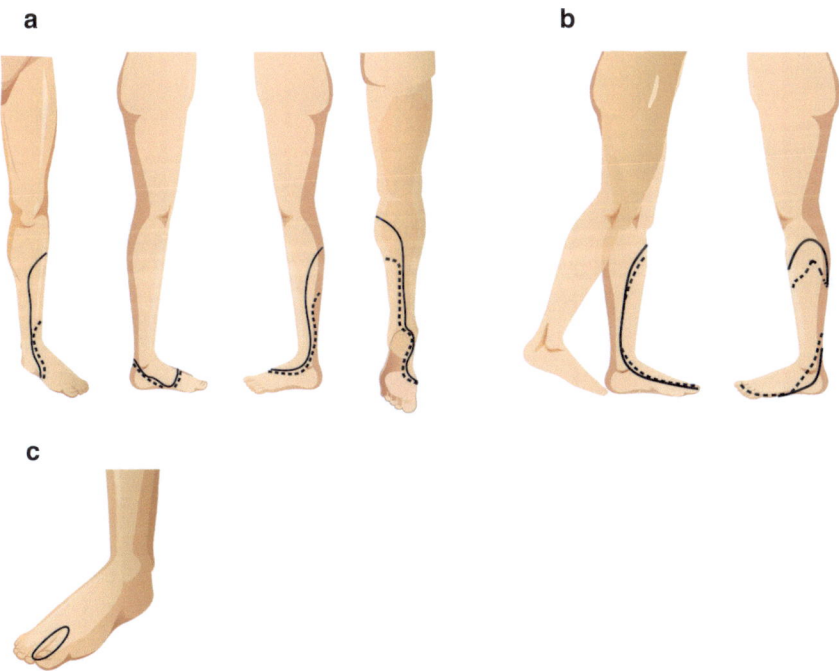

Fig. 1.9 The approximate area within which sensory changes may be found in lesions of (**a**) sciatic nerve below the branch of the posterior cutaneous nerve of the thigh, anterior, medial, lateral and posterior view; (**b**) common peroneal nerve above the origin of the superficial peroneal nerve, medial and lateral view; (**c**) deep peroneal nerve

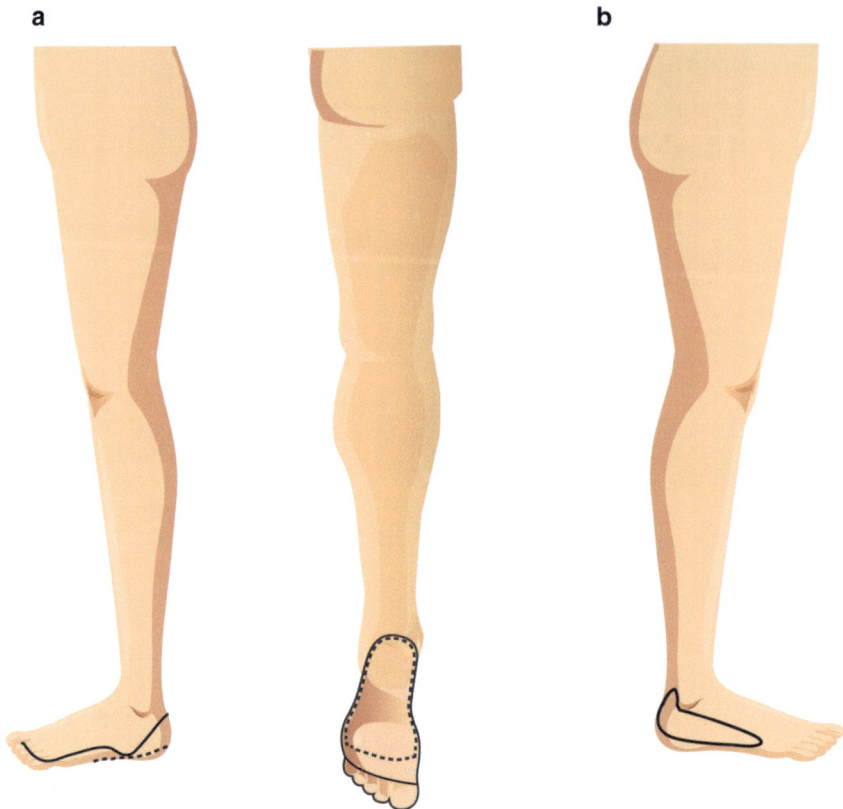

Fig. 1.10 The approximate area within which sensory changes may be found in lesions of (**a**) tibial nerve, lateral and posterior view; (**b**) sural nerve, lateral view

In localizing a peripheral neuropathy, it should be kept in mind that the area of skin innervation by peripheral nerves also varies considerably. Moreover, nerve fascicules supplying individual muscles or cutaneous areas are already definite proximally in the plexus, and partial nerve lesions involving only certain fascicles may mimic a more distal lesion of the nerve or even of one of its branches.

1.4 Tendon Reflexes

Tendon reflexes are elicited by a fast blow on the tendon of a muscle. The resulting muscle stretching and vibration excite the muscle spindles, generating impulses that, through the Ia afferent fibers, travel to the spinal cord and excite the α-motor neurons of the stimulated muscle. Tendon reflexes can be reduced or absent in PNS disorders because of the involvement of Ia fibers, efferent fibers from α-motor neuron to the muscle, and the muscle itself. Generalized areflexia is a characteristic sign

Table 1.2 Tendon reflex, muscle, nerve, and root innervation

Reflex	Muscle	Nerve	Roots
Biceps	Biceps brachialis	Musculocutaneous	C5, C6
Brachioradialis	Brachioradialis	Radial	C5, C6
Triceps	Triceps brachii	Radial	C6, C7, C8
Finger flexors	Finger flexors	Median and ulnar	C7, C8, T1
Knee	Quadriceps femoralis	Femoral	L2, L3, L4
Ankle	Triceps surae	Sciatic/tibial	S1, S2

of a polyneuropathy. Localized hypo-areflexia may be present, even without muscle weakness or sensory loss, in cervical and lumbar radiculopathies and contributes to locating the disorder (Table 1.2). On the other hand, hyperactive reflexes will point to upper motor neuron involvement as in amyotrophic lateral sclerosis or cervical myelopathy.

References

1. Aids to the examination of the peripheral nervous system. London: Baillière Tindal; 1986.
2. Stewart JD. Focal peripheral neuropathies. 2nd ed. New York: Raven Press; 1993.
3. Wilbourn AJ. Radiculopathies. In: Brown WF, Bolton CF, editors. Clinical electromyography. 2nd ed. Boston: Butterworth; 1993.

Part II

A Guide to Routine Nerve Conduction Studies

Basic Principles of Nerve Conduction Studies

2

Following are reported some very basic concepts on nerve conduction studies. Superficial peripheral nerves can be easily stimulated with a brief electrical pulse applied to the skin. Motor, sensory, or mixed nerve studies can be performed by stimulating the nerve and placing the recording electrodes over a muscle, a cutaneous sensory nerve, or the entire mixed nerve, respectively.

2.1 Potential Recording

Potentials recorded during nerve conduction studies are the result of the difference between the signals recorded at the active (A) and reference (R) electrodes that are amplified and then displayed on the electromyography (EMG) machine screen. A factor influencing the recording in nerve conduction studies, above all when recording small amplitude potentials, is the electrical noise interference (50 Hz in Europe and 60 Hz in the USA) generated mainly by ambient electrical devices. If the same electrical noise is present at both A and R, it will be subtracted, and only the signal of interest will be amplified (common mode rejection). To attain identical electrical noise at A and R, the impedance at each electrode should be the same. If the impedance is different (impedance mismatch), the electrical noise will induce a different voltage at each electrode input that will be amplified and displayed, obscuring the signal. To reduce electrode impedance mismatch, (1) check that A and R are of the same type, (2) ensure that all contacts are intact without any worn or broken connections, (3) scrub and clean the skin using sandpaper and alcohol, (4) apply conducting electrode jelly between the skin and the electrodes, (5) secure the electrodes firmly to the skin with tape, and (6) use coaxial recording cables. Electrical noise can be eliminated by filtering the signal. Therefore, potentials recorded during nerve conduction studies and needle EMG pass through a low- and a high-frequency filter before being displayed. The role of the filters is to reproduce the signal of interest while excluding both low- and high-frequency noise. Low-frequency filters, also called high-pass filters, exclude signals below a set frequency while allowing higher-frequency signals to pass. High-frequency filters, also called low-pass filters, exclude signals above a certain frequency while allowing lower-frequency signals to pass. By allowing the signal to pass through a certain "passband," some noise can be excluded. Different filter bandwidths are used in the laboratories around the world; however, for motor conduction studies, the low- and high-frequency filters are usually set at 10 Hz and 10 kHz and for sensory conduction studies at 20 Hz and 2 kHz. Changing filter bandwidth has an effect on compound muscle action potential (CMAP) and sensory nerve action potential (SNAP) parameters; therefore, conduction studies should be done with standardized filter settings and results compared with control values obtained using the same filter settings.

2.2 Nerve Stimulation

For nerve stimulation, a constant current stimulator, available in most recent EMG equipment, is recommended. This stimulator adjusts the voltage necessary for the set current depending on the impedance between the electrode and the skin. On the other hand, constant voltage stimulators, which generate a specified voltage, have the drawback that the actual stimulus current delivered depends on the impedance of the skin. In the stimulator, the cathode is the terminal where current flows out (indicated in the manual in black), whereas the anode is the terminal where current flows in (indicated in red). Depolarization of a nerve occurs under the cathode of the stimulator, and the depolarization proceeds in both directions along the nerve.

2.3 Motor Conduction Studies

Motor conduction studies are technically less demanding than sensory and mixed nerve studies and usually are performed first. This allows the examiner to know where the nerve runs, where it should be stimulated, and how much current is needed to stimulate the nerve. The CMAP is the summation of individual muscle fiber action potentials, and many types of electrodes are used for recording. The silver/silver chloride 1 cm diameter electrode has a recording area of 0.79 cm^2, and it should be underlined that most of the control values for the motor conduction studies techniques reported in this manual have been obtained with these electrodes. Currently, self-adhesive disposable electrodes of different sizes are increasingly used. Large-size electrodes reduce the variations of CMAP morphology and amplitude that can occur with different placements of small electrodes. However, it should be reminded that with larger electrodes, the CMAP amplitude and area actually decrease because more motor unit potentials are recorded resulting in a greater cancellation among positive and negative phases of the individual potentials. Therefore, different sizes of electrodes require specific control values for the CMAP amplitude and area, and it is recommended to always use, in follow-up studies, electrodes of

2.3 Motor Conduction Studies

the same size. For motor conduction studies, the recording electrode A is placed on the muscle belly (over the motor end plate), and R is placed usually over the muscle tendon. In the figures of this manual, A is connected to a black wire, and R is connected to a red wire. The ground electrode should be placed always between the stimulating and recording electrodes to reduce the stimulus artifact and for electrical safety reasons. The stimulator is placed over the nerve that supplies the muscle, with the cathode closest to the active recording electrode. It is helpful to remember the memo "black to black," indicating that the black electrode of the stimulator (the cathode) should be facing the black recording electrode (the active electrode). For motor conduction studies, the gain usually is set at 2–5 mV per division, and the analysis time at 2–5 ms per division. The duration of the electrical stimulus is usually 0.2 ms. The current intensity is slowly increased from 0 with 5–10 mA increments. When the current is increased to the point that the CMAP no longer increases in amplitude, it is assumed that all nerve fibers have been excited and that maximal stimulation has been achieved. The current is then increased by another 10–20% to ensure supramaximal stimulation. Most normal nerves require a current intensity in the range of 20–50 mA to achieve supramaximal stimulation. The CMAP is a biphasic potential with an initial negativity (upward deflection from the baseline). If the A electrode is not correctly placed in the end plate zone, the CMAP has an initial positive deflection, and the electrode should be repositioned. An initial positive deflection may also be caused by the recording of a volume-conducted CMAP generated by adjacent muscles innervated by the same stimulated nerve. This may occur in radial, ulnar, and tibial nerve studies, and it is not always possible to eliminate the initial positivity.

For each stimulation site, the latency, amplitude, duration, and area of the CMAP are measured (Fig. 2.1). The CMAP amplitude is measured in millivolts and is most commonly measured from the baseline to the negative peak. The amplitude may be also measured from the first negative peak to the next positive peak, but, in this case, the amplitude is variably influenced by the position of the reference electrode.

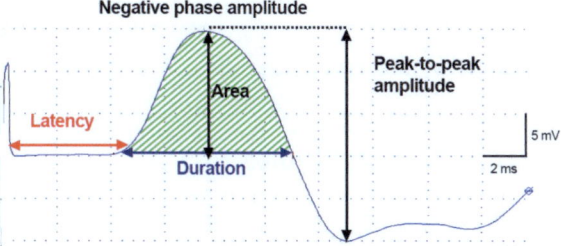

Fig. 2.1 Compound muscle action potential parameters and its measurements. Note that negative is up and positive is down

The CMAP amplitude reflects the number of activated axons and muscle fibers and the synchronization among conduction velocity (CV) of nerve fibers. The CMAP duration is measured in milliseconds and is defined as the time from the onset of the initial negative phase to the return to the baseline. This is preferred to the measurement of the total duration including the terminal positive phase because the precise return to the baseline is often not easily determinable. The CMAP duration reflects the synchronization of the CV of motor nerve fibers. In dispersed CMAP, the duration is conventionally measured from the onset of the first negative phase to the return to the baseline of the last negative deflection. The CMAP area is expressed in millivolts per millisecond, and the area of the negative peak is measured conventionally. The CMAP area cannot be determined manually, but the calculation is readily performed by modern EMG machines. The negative peak CMAP area is another measure reflecting the number of nerve and muscle fibers that depolarize and their synchronization. The distal CMAP latency is the time from the stimulus to the initial deflection from the baseline. The latency is expressed in milliseconds and reflects only the fastest-conducting largest motor fibers, whereas the slower-conducting fibers contribute to the CMAP amplitude area and duration. If the electrodes are not over the end plates, the CMAP has a positive rather than negative onset, and the latency to the positive peak includes also the time for conduction along the muscle fiber to the recording electrode. It is debated where to measure the onset latency of a CMAP with a positive onset; however, independent of whether it is measured at the onset or at the peak of the positive deflection, it should be kept in mind that in this case the measurement of distal latency is not completely reliable for assessing a distal nerve pathology.

Motor nerve CV cannot be calculated by performing only a single distal stimulation because the distal motor latency (DML) includes: the conduction time along the distal motor axon to the neuromuscular junction, the neuromuscular junction transmission time, and the depolarization time of muscle fibers. Therefore, to calculate a motor CV, two stimulation sites must be used, one distal (i.e., for the median nerve at the wrist) and one proximal (at the antecubital fossa). The motor CV is then calculated by dividing the distance between the two stimulation sites (in millimeters) by the latency difference of CMAPs from the two stimulation sites (in milliseconds). In the case of CMAPs with a positive onset, the CV can be calculated paying attention to the fact that the onset latency is measured at the same point of CMAPs from distal and proximal stimulation.

2.4 Sensory Conduction Studies

When a nerve is electrically stimulated, the impulse is conducted in both directions away from the stimulation site. In sensory nerve conduction studies, when stimulating the skin distally and recording over the nerve proximally, the technique is termed

2.4 Sensory Conduction Studies

orthodromic (which is the physiological path of sensory impulse conduction). Under the reverse condition, the technique is called antidromic. For both techniques, always remember that the A electrode should be closer to the cathode of the stimulator (black-to-black rule). The gain is set usually at 10–20 µV per division, and an electrical pulse of 0.1 or 0.2 ms duration is used for stimulation. Most normal sensory nerves require a current intensity of 5–30 mA to achieve supramaximal stimulation. The sensory nerve action potential (SNAP) is the summation of the individual sensory fiber action potentials. Orthodromic SNAPs are usually triphasic with an initial positive deflection. In antidromic recordings, the SNAP is usually biphasic. The SNAP amplitude is measured peak to peak or from the baseline to the negative peak (Fig. 2.2).

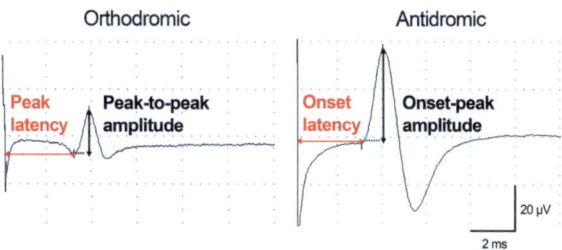

Fig. 2.2 Median SNAP recorded orthodromically (digit 3, wrist) and antidromically (wrist, digit 3) in the same subject. Note the higher amplitude of the antidromic SNAP

The SNAP latency is measured from the stimulus to the initial positive peak for triphasic SNAP or to the initial negative deflection for biphasic SNAP and represents nerve conduction time from the stimulus site to the recording electrodes for the largest and fastest cutaneous sensory fibers. Some electroneuromyographers measure the latency of SNAP at the negative peak because it is sometimes easier to detect. However, be aware that this measurement does not reflect the conduction of the largest-diameter fast-conducting fibers.

In this manual, mostly antidromic techniques are shown because they are technically easier and yield higher amplitude SNAPs. Only one stimulation site is required to calculate a sensory CV according to the simple formula of distance (in millimeters) divided by latency (in milliseconds).

Recording small potentials as SNAPs can be hampered by the electrical noise that can be reduced by electronic averaging through which potentials obtained with serial stimulations are digitized and mathematically averaged. Because electrical noise is random, positive and negative phases of electrical noise will cancel each other, as more recordings are averaged, leaving the potential of interest. Electronic averaging is also helpful in stabilizing the electrical baseline so that the onset latency and amplitude of SNAP can be more correctly measured (Fig. 2.3).

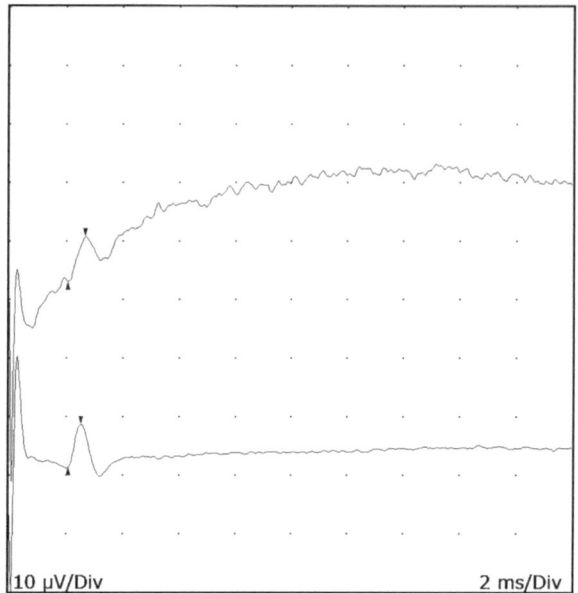

Fig. 2.3 Sural nerve conduction. Upper trace is the single SNAP, and lower trace is the average of eight trials

2.5 Temperature and Nerve Conduction

Temperature is the most important factor in influencing nerve conduction. Low cutaneous temperature (below 32 °C) prolongs DML of 0.2 ms per °C and reduces CV of 1.5–2.5 m/s per °C, whereas it increases amplitudes of the SNAP more than the CMAP. If the reduced limb temperature is not taken into consideration, slow nerve CVs may be misinterpreted as the expression of pathology. A common mistake is to diagnose a polyneuropathy on the basis of diffused slightly slow CVs and prolonged DMLs simply due to cold limb temperature. Another common error is that prolonged distal median motor and sensory latencies in a cool limb may be misdiagnosed as a median neuropathy at the wrist (carpal tunnel syndrome). Distal limb temperatures should be routinely checked in all patients and ideally maintained between 32 and 34 °C by a heating lamp, warm packs, or a hot tub. Keep in mind that when a limb is warmed, the skin temperature usually reaches the desired value several minutes before the underlying nerve and muscle. For very cool limbs (~28 °C), it may require 20–40 min for the underlying nerve temperature to equilibrate. If limb warming is not possible or is difficult to achieve (i.e., studies in the intensive care unit), apply conversion factors of 1.5–2.5 m/s per °C for the conduction velocity and 0.2 ms per °C for DML. Modern EMG equipment can monitor limb temperature and automatically report corrected values for the conduction velocity and latency. However, these correction factors are derived primarily from individuals with normal nerves and may not be correct for diseased nerves.

2.6 Measurement of Distance

To calculate CV, the distance along the nerve between different stimulation points must be correctly measured. It is usually assumed that the surface distance represents the true underlying length of the nerve. However, there are exceptions, the most important being the ulnar nerve that is slack and redundant across the elbow when the arm is in the extended position. If surface distance measurement is made with the extended arm, the true length of the nerve is underestimated factitiously slowing the CV. The optimal position for measuring the distance is with the elbow flexed between 70° and 90°. Surface distance measurements of the radial nerve as it spirals around the humerus and of the median and ulnar nerves between the axilla and Erb's point often are inaccurate. In these situations, an obstetric caliper should be used to more precisely assess the true length of the underlying nerve. To ensure correct measurement of the distance, (1) mark the site of the cathode with a pen before the stimulator is removed from the skin; (2) hold the tape along the anatomical course of the nerve during measurement; (3) use a caliper for proximal nerve stimulation (brachial plexus); (5) repeat the measurement if an unexpected CV value is obtained.

2.7 Reference Values

In nerve conductions, the values of the measured parameters are compared with reference values to determine whether the results are normal or not. Ideally, each laboratory should develop its own set of control values, but this often does not happen. In this manual, for each conduction study technique, the control values and the upper limit of normal (ULN) or lower limit of normal (LLN) are reported that can be used for reference as long as the same technique is employed. It should be underlined that normative data for nerve conduction studies do not always follow a Gaussian distribution, and adopting, as the range of normality, the mean ± 2 SD, as often done, may be incorrect because data should be converted by a logarithmic square root. An acceptable substitute is to use, as the normal range, the 97th and 3rd percentile of the observed values.

Cranial Nerve Studies

3.1 Facial Nerve Study

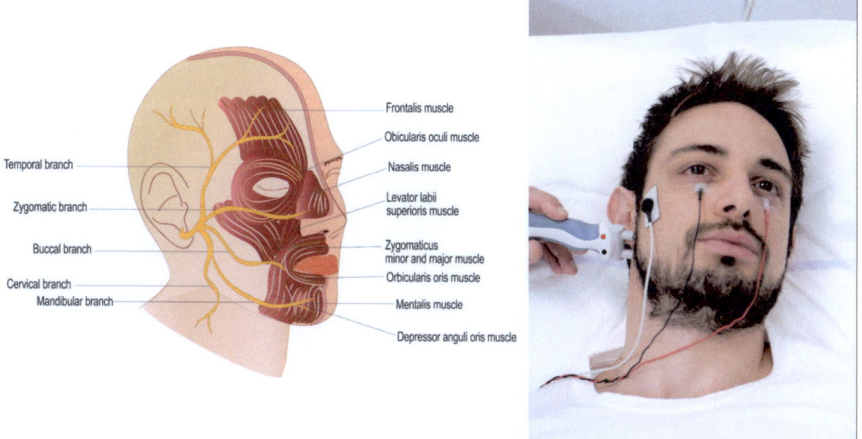

Fig. 3.1 Left: the facial nerve, its major branches and the supplied muscles. Right: Facial neve study with stimulation at the stylomastoid foramen and recording from orbicularis oculi muscle

Nerve Fibers Tested and Route
Motor fibers of the facial nerve (Fig. 3.1, left).

Recording Site
Orbicularis oculi muscle. The active electrode (A) is placed under the eye at midline, and the reference electrode (R) is placed on the contralateral muscle. G is placed on the cheek (Fig. 3.1, right).

Other facial muscles such as the frontalis, nasalis, and mentalis can be used as recording sites using similar montages (A is placed over the muscle, R is placed on the contralateral muscle).

Concentric needle recording from orbicularis oris has been also described.

Stimulation Site

Below the mastoid and behind the mandible with the cathode over the stylomastoid foramen. Distance: variable (Fig. 3.1, right).

Control Values

83 subjects [1].

Latency (ms), 2.9 ± 0.4; the upper limit of normal (ULN) (mean + 3SD), 4.1. ULN side-to-side difference: 0.6 ms (Fig. 3.2).

Concentric needle recording from orbicularis oris [2].

Latency (ms) to the onset of negative CMAP deflection: 4.0 ± 0.5.

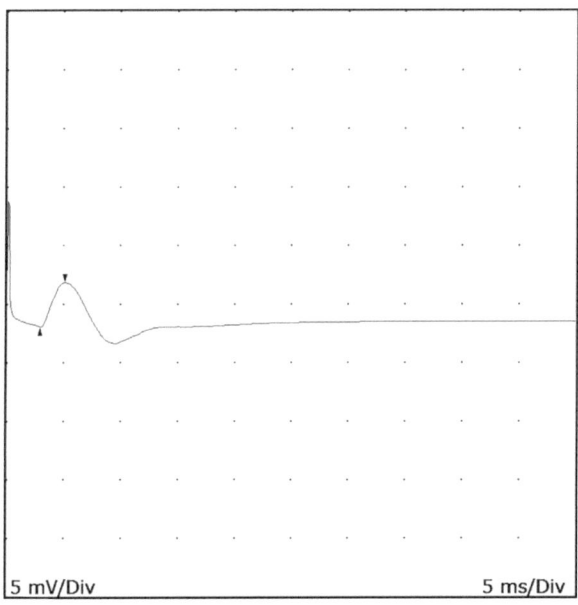

Fig. 3.2 Facial nerve study recording from orbicularis oculi, control subject: the DML is 3.1 ms, and the CMAP amplitude is 3.9 mV

Comments

- With this technique, the whole facial nerve at the stylomastoid foramen, where it exits from the skull, is stimulated.
- Often high stimulation intensities are required, and the study can result in discomfort.
- With high intensities of stimulation, the masseter muscle can be directly activated, and a volume-conducted response can be recorded with surface electrodes

from the orbicularis oculi obscuring the response or, in the case of facial nerve lesion, fallaciously suggesting a good prognosis. In this situation, the observation of the face reveals the contraction of the masseter.
- Individual facial branch stimulation is often easier and more comfortable for the patient (see facial motor branches study).
- The CMAP amplitude varies substantially inter-individually, and a surface-recorded CMAP ≥1 mV is generally considered normal. The comparison between the two sides in the same individual is more significant than the absolute value, and a side-to-side difference of CMAP amplitude greater than 50% is usually considered abnormal.
- Serial recordings of CMAP measuring amplitude can be useful to predict the prognosis in facial nerve injury by documenting the degree of axonal loss. The distal axons remain excitable for a few days even after complete proximal transection, but excitability is usually lost after 1 week in parallel with the progression of axonal degeneration.
- This study is useful in assessing facial nerve palsies and demyelinating neuropathies.

3.2 Facial Branches Study

3.2.1 Frontal Branch

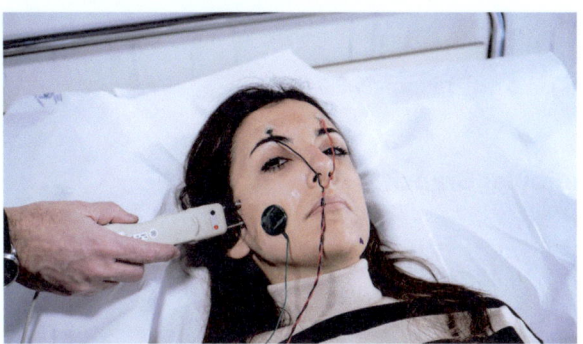

Fig. 3.3 Facial nerve study: frontal branch

Recording Site
Frontalis muscle. A is placed above the eyebrow, slightly medial to the center. R is placed on the contralateral frontalis. G is placed on the cheek (Fig. 3.3).

Stimulation Site
Three to four fingerbreadths lateral to the eye. Distance: variable (Fig. 3.3).

3.2.2 Zygomatic Branch

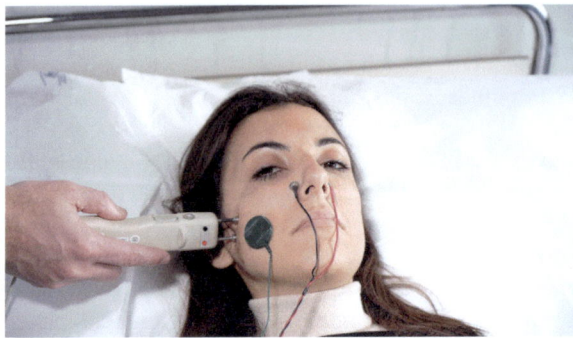

Fig. 3.4 Facial nerve study: zygomatic branch

Recording Site
Nasalis muscle. A is placed just lateral to the mid-nose, and R is placed on the contralateral side of the nose at the same location. G is placed on the cheek (Fig. 3.4).

Stimulation Site
Over the zygomatic bone just anterior to the ear. Distance: variable (Fig. 3.4).

Control Values
44 subjects [3].
　Latency (ms), 3.57±0.35; range, 2.8–4.1.

3.2.3 Mandibular Branch

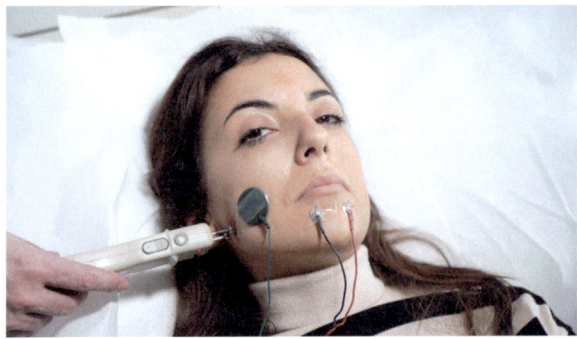

Fig. 3.5 Facial nerve study: mandibular branch

Recording Site
Mentalis muscle. A is placed over the chin, and R is placed over the contralateral mentalis. G is placed on the cheek (Fig. 3.5).

Stimulation Site
Over the angle of the jaw. Distance: variable (Fig. 3.5).

3.2.4 Facial Branches Stimulation in Studying the Lateral Spread of Excitation

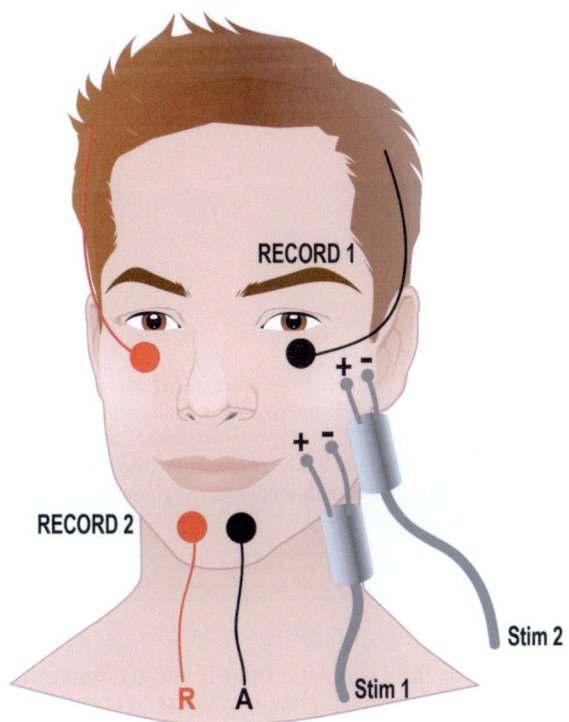

Fig. 3.6 Facial branches stimulation in studying lateral spread of excitation

Recording Sites
Dual-channel simultaneous recording from the orbicularis oculi muscle (record 1) and mentalis muscles (record 2) (Fig. 3.6).

Stimulation Sites
Zygomatic and mandibular branches as shown in the facial branches study (Fig. 3.6).

In normal subjects, only the muscle innervated by the stimulated branch shows a response. In patients with hemifacial spasm, a delayed response is also seen in the muscle innervated by the branch not being directly stimulated. This is because when a facial nerve branch is stimulated, the depolarization travels both orthodromically and antidromically. The antidromic volley traveling to the site of nerve injury or compression presumably spreads ephaptically to nerve fibers of contiguous branches, resulting in a response in muscles innervated by those branches [4] (Fig. 3.7).

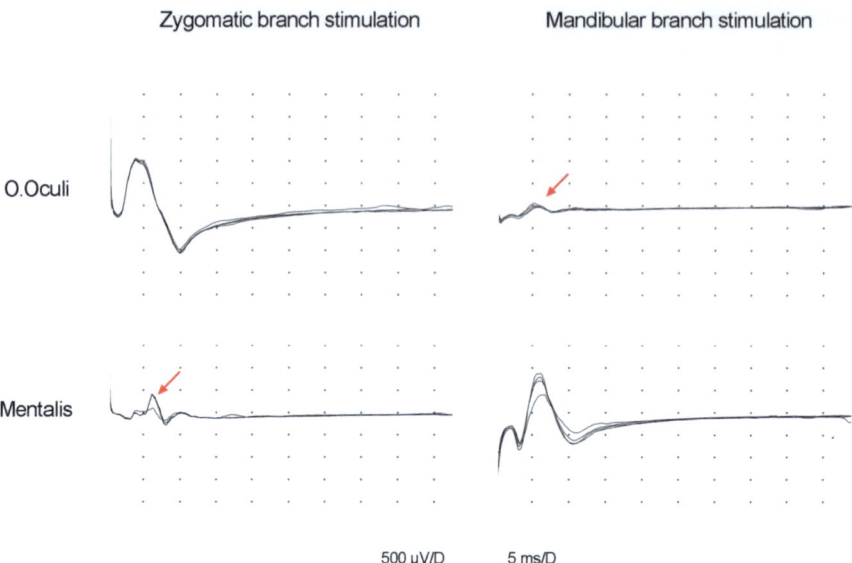

Fig. 3.7 Study of lateral spread of excitation. Selective facial nerve branches (zygomatic and mandibular) stimulation and co-recording from orbicularis oculi and mentalis muscles (superimposed tracings) in a patient with hemifacial spasm after facial palsy. Stimulation of the zygomatic branch produces an abnormal small amplitude response in the mentalis and stimulation of the mandibular branch induces an abnormal small amplitude response in the orbicularis oculi (red arrows) as evidence of ephaptic transmission

Comments

- Selective stimulation of the zygomatic branch and recording from the nasalis muscles usually provides a well-defined CMAP avoiding the recording of volume-conducted responses from the masseter.
- Selective stimulation of facial nerve branches is used in studying the lateral spread of excitation in hemifacial spasm to demonstrate ephaptic transmission (Figs. 3.6 and 3.7).

3.3 Blink Reflex Study

Fig. 3.8 Blink reflex study

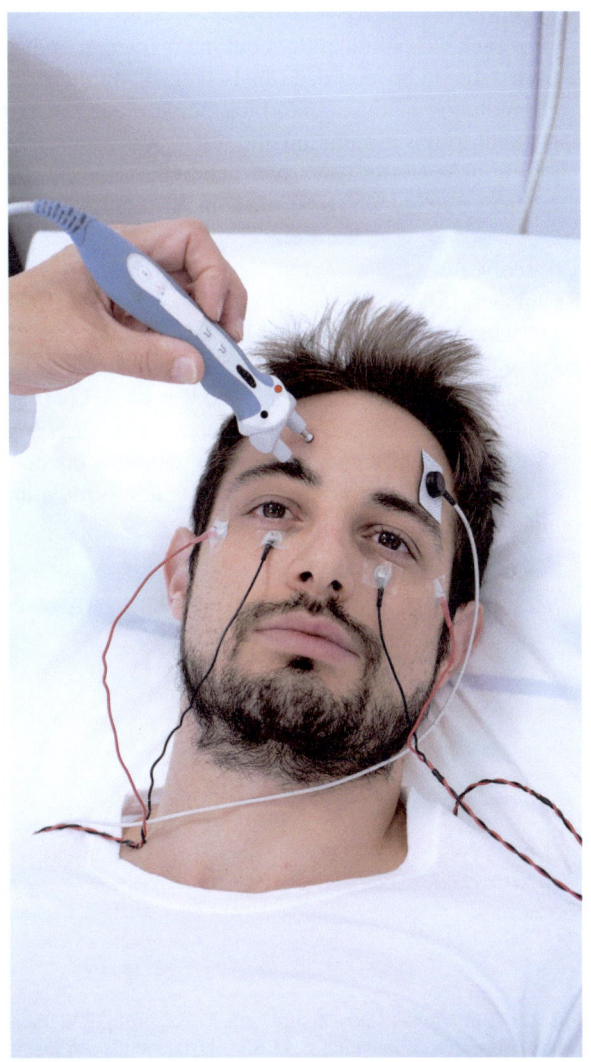

Nerve Fibers Tested and Route
Afferent trigeminal sensory fibers, efferent facial motor fibers, and their central connections in the pons and medulla.

Recording Site
A is placed over the orbicularis oculi bilaterally, and R is placed at the lateral angle of the eye. G is placed on the forehead (Fig. 3.8).

Stimulation Site

The cathode is over the supraorbital nerve at the supraorbital notch, and the anode is superior (Fig. 3.8).

For each side stimulation, both the ipsilateral and contralateral orbicularis oculi muscles should be recorded (dual-channel simultaneous recording).

Stimulation of the supraorbital nerve elicits an early R1 response on the side ipsilateral to the stimulation and a late R2 response bilaterally (Fig. 3.9). R1 is a biphasic or triphasic wave that habituates slowly, whereas R2 is polyphasic and habituates quickly with a decrease in the amplitude and duration.

Control Values

83 subjects [5].

R1 latency (ms), 10.5 ± 0.8; ULN (mean + 3 SD), 13.

Ipsilateral R2 latency (ms), 30.5 ± 3.4; ULN (mean + 3 SD), 40.

Contralateral R2 latency (ms), 30.5 ± 4.4; ULN (mean + 3 SD), 41.

Side-to-side difference in R1 latency (ms): ULN (mean + 3 SD), 1.2.

Side-to-side difference in R2 (ms) with dual-channel simultaneous recording: 5.

Side-to-side difference in R2 (ms) by first stimulating one side and then the other (single-channel recording): 7.

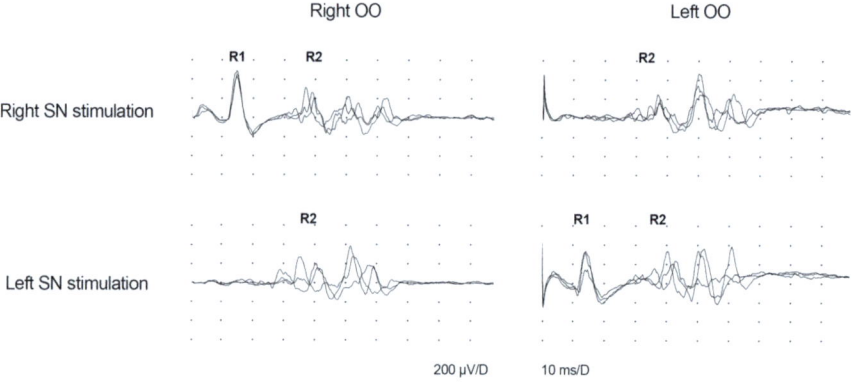

Fig. 3.9 Blink reflex study, control subject, superimposed tracings. SN, supraorbital nerve; OO, orbicularis oculi. Stimulating right SN: R1 latency is 9.6 ms, ipsilateral R2 latency is 33.5 ms, contralateral R2 latency is 30.3 ms. Stimulating left SN: R1 latency is 10.7 ms, ipsilateral R2 latency is 33.8 ms, contralateral R2 latency is 29 ms

3.3 Blink Reflex Study

Comments

- R1 latency reflects impulse conduction along the large myelinated (Aβ) primary afferents from the ophthalmic division to the main sensory nucleus of the trigeminal nerve in the pons and then, through a disynaptic pathway, to the facial nerve nucleus and along the efferent pathway of the ipsilateral facial nerve. R2 latency reflects impulse conduction along the same afferent pathway of R1 to the nucleus of the spinal tract of trigeminus and, through multiple synapses in the pons and lateral medulla, to both the ipsilateral and contralateral facial nerve nuclei and along the efferent pathways of the facial nerves bilaterally.
- During the test, the subject should be relaxed, lying supine with the eyes either open or gently closed.
- Supramaximal stimulation can be achieved with low intensities, typically 10–15 mA.
- Lateral rotation of the anode of the stimulator around the cathode helps to reduce the stimulus artifact.
- The anode of the stimulator should be away from the midline to avoid the stimulation of the contralateral supraorbital nerve resulting in the recording of a contralateral R1 (Fig. 3.10) [6].

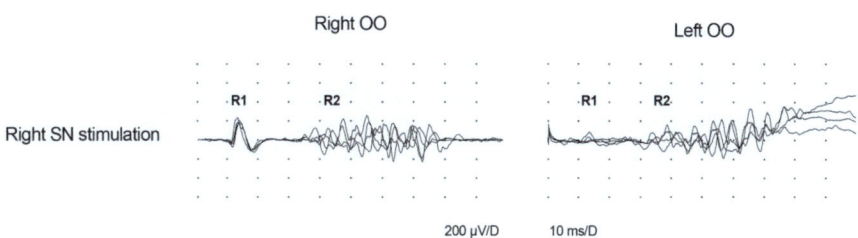

Fig. 3.10 Blink reflex study, control subject, superimposed tracings. SN, supraorbital nerve; OO, orbicularis oculi. Stimulating the right SN a small amplitude R1 is recorded also from the left OO. This is due to spreading of the stimulus to the left SN and can be avoided by rotating laterally the anode of the stimulator

- To prevent habituation, stimulate at several-second intervals.
- Usually, four to six traces are recorded and superimposed to determine the minimal R1 and R2 latencies.
- Latencies of R1 and R2 should be measured to the first deflection from the baseline.
- This study is useful in assessing facial nerve palsies, demyelinating neuropathies, and brainstem lesions.

3.4 Spinal Accessory Nerve Study

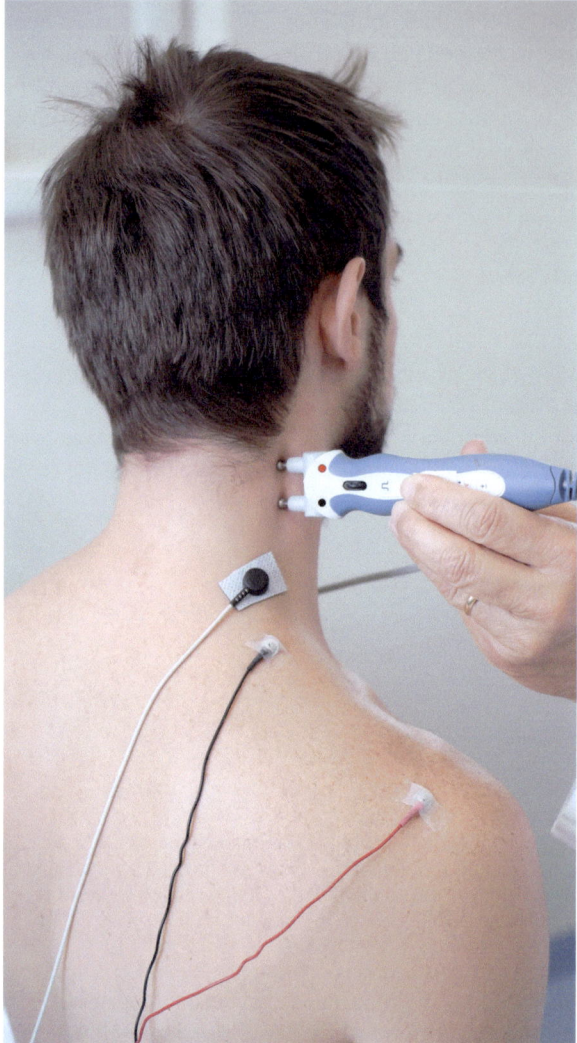

Fig. 3.11 Spinal accessory nerve study

Nerve Fibers Tested and Route
Spinal accessory nucleus (C1–C5), spinal accessory nerve. The spinal roots of the accessory nerve emerge from the lateral aspect of the cord and ascend to enter the skull through the foramen magnum. The nerve leaves the skull through the jugular foramen to innervate the sternocleidomastoid and the trapezius muscles.

3.4 Spinal Accessory Nerve Study

Recording Site

A is placed over the upper trapezius about 9 cm lateral to the 7th cervical spinous process, and R is placed on the acromion. G is placed between the stimulating and recording electrodes (Fig. 3.11).

Stimulation Site

The cathode is placed 1–2 cm to the posterior border and slightly above the midpoint of the sternocleidomastoid muscle. The anode is superior (Fig. 3.11).

Control Values

28 subjects [7].
  Latency (ms), 2.3 ± 0.4; range, 1.7–3.0.
  Peak-to-peak amplitude (mV): >3 (Fig. 3.12).

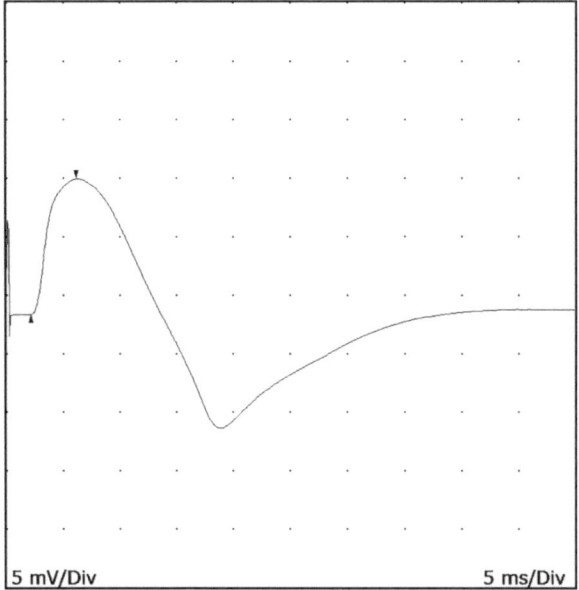

Fig. 3.12 Spinal accessory nerve study, control subject: DML is 2.2 ms, CMAP amplitude (negative peak) is 11.5 mV

Comments

- The shoulder should shrug when the spinal accessory nerve is stimulated.
- Care should be taken to avoid stimulation of the brachial plexus.
- The commonest cause of accessory nerve injury is biopsy of the lymph nodes in the posterior triangle of the neck. Nerve damage may be unavoidable in radical neck dissection for cancer.
- Weakness of the upper and middle part of the trapezius induces winging of the scapula.
- If the nerve is damaged proximally, wasting and weakness of the sternocleidomastoid muscle are associated.

References

Facial Branches Study

1. Kimura J, Powers JM, Van Allen MW. Reflex response of orbicularis oculi muscle to supraorbital nerve stimulation. Arch Neurol. 1969;21:193–9.
2. Taylor N, Jebsen RH, Tenckoff HA. Facial nerve conduction latency in chronic renal insufficiency. Arch Phys Med Rehabil. 1970;51:259–63.
3. Ma DM, Liveson JA. Nerve conduction handbook. Philadelphia: FA Davis; 1983.
4. Valls-Solé J. Electrodiagnostic studies of the facial nerve in peripheral facial palsy and hemifacial spasm. Muscle Nerve. 2007;36:14–20.

Blink Reflex Study

5. Kimura J, Powers JM, Van Allen MW. Reflex response of orbicularis oculi muscle to supraorbital nerve stimulation. Arch Neurol. 1969;21:193–9.
6. Soliven B, Meer J, Uncini A, Petajan J, Lovelace R. Physiologic and anatomic basis for contralateral R1 in blink reflex. Muscle Nerve. 1988;11:848–51.

Spinal Accessory Nerve Study

7. Ma DM, Liveson JA. Nerve conduction handbook. Philadelphia: FA Davis; 1983.

4 Cervical, Brachial, and Upper Limb Nerve Studies

4.1 Phrenic Nerve Study

Fig. 4.1 Phrenic nerve study

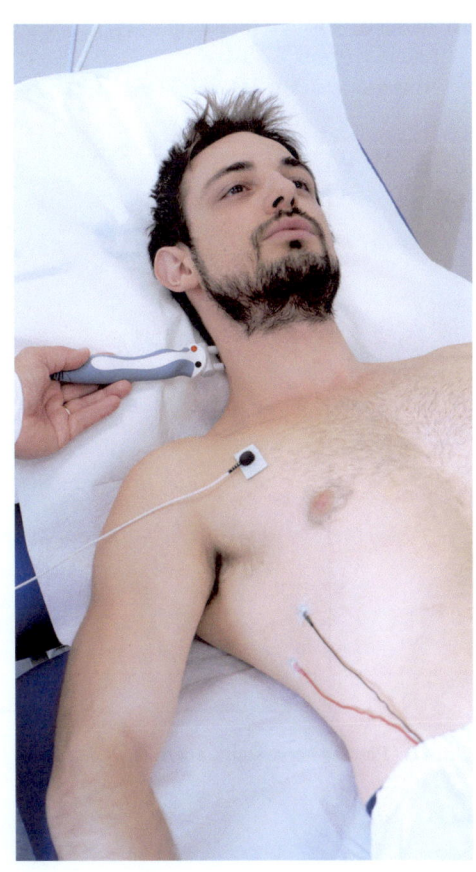

Nerve Fibers Tested and Route
Motor fibers of C3, C4, and C5 roots, phrenic nerve.

Recording Site
The active electrode (A) is placed on the eighth intercostal space in the anterior axillary line; the reference electrode (R) is placed 3.5–5 cm posteriorly on the eighth or ninth intercostal space. The ground electrode (G) is placed in the upper chest between the stimulating and recording electrodes (Fig. 4.1).

Stimulation Site
With the subject supine and the neck in a neutral position or slightly extended, stimulation is applied at the posterior border of the sternocleidomastoid (SCM) muscle at the level of the thyroid cartilage; the anode is superior to the cathode (Fig. 4.1).

Control Values
18 subjects, 4 tested bilaterally [1].
  Latency (ms): 7.7 ± 0.8.
  Amplitude (µV): 160–500 (Fig. 4.2).

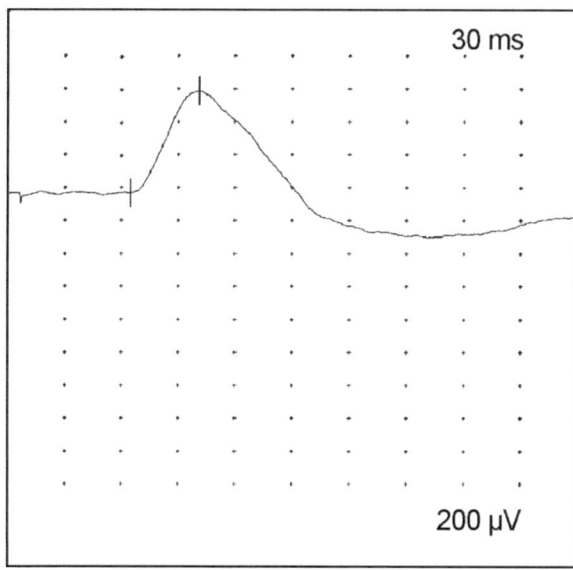

Fig. 4.2 Phrenic nerve study, control subject: DML is 6.9 ms, CMAP amplitude is 0.6 mV

4.1 Phrenic Nerve Study

Comments

- To localize the stimulation site, both heads of the SCM muscle can be easily localized by asking the patient to turn the neck.
- During stimulation, hold the stimulator in place and apply firm pressure.
- The spinal accessory nerve can be also stimulated (causing contraction of the trapezius).
- If the stimulator is not in the correct location, the brachial plexus can be stimulated causing movement of the shoulder and arm and recording of a volume-conducted compound muscle action potential (CMAP) from other muscles with shorter latency, low amplitude, and initial positivity.
- In thin individuals, the diaphragm contraction often can be seen, and it is similar to a hiccup.
- The phrenic nerve is difficult to study in obese individuals.
- Both sides should be examined.
- Because of amplitude variability, it is helpful to repeat the recording several times to obtain for each side the highest CMAP amplitude.
- CMAP amplitude is slightly greater during inspiration, and deep breathing should be avoided during the test.
- Do not perform this study in the intensive care unit in patients who have an external pacemaker (risk of current spread to the heart); caution is needed in the presence of a central catheter, implanted cardiac pacemaker, or cardioverter-defibrillator.
- The phrenic nerve can be injured in the neck by wounds and during jugular vein catheterization and can be involved by metastases, usually from breast cancer. In the chest, the nerve may be infiltrated by intrathoracic malignant tumors, and the left phrenic can be involved in open heart surgery.
- The motoneurons of the phrenic nerve can be involved in amyotrophic lateral sclerosis, and occasionally, respiratory insufficiency may be the presenting feature. The phrenic nerve may be involved in Guillain-Barré syndrome and, even solely, in neuralgic amyotrophy (acute brachial plexitis).
- It should be underlined that unilateral phrenic nerve involvement is usually asymptomatic or induces only mild dyspnea.

4.2 Long Thoracic Nerve Study

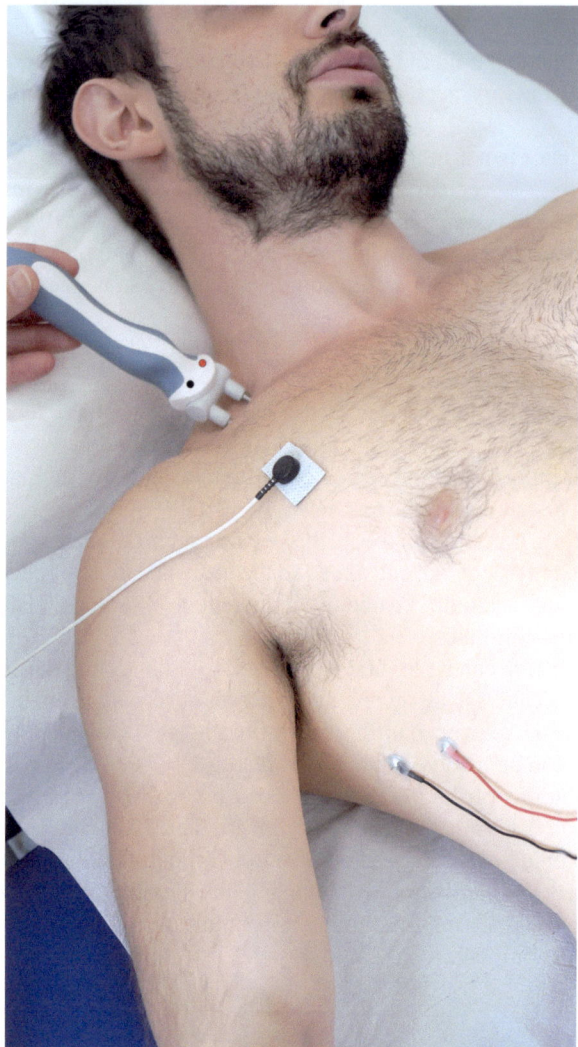

Fig. 4.3 Long thoracic nerve study with surface electrode recording

4.2 Long Thoracic Nerve Study

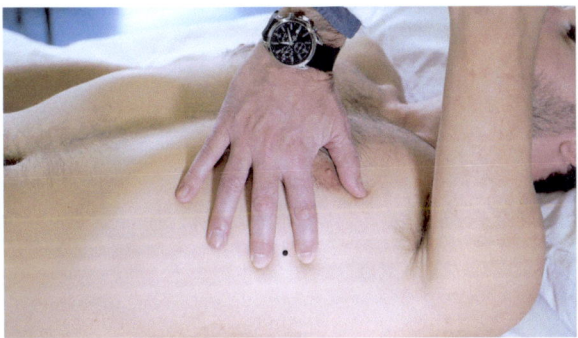

Fig. 4.4 Long thoracic nerve study with concentric needle recording. The insertion point of the needle electrode is shown

Nerve Fibers Tested and Route
C5, C6, and C7 roots, before the brachial plexus.

Recording Site
Surface recording with A over the serratus anterior muscle along the midaxillary line on the 5th rib; R is placed 3 cm anterior to A. G is placed between the stimulating cathode and A. For concentric needle recording, the needle electrode is placed at the digitation of the serratus anterior along the midaxillary line over the 5th rib (Figs. 4.3 and 4.4).

Stimulation Site
At Erb's point, slightly above the clavicle and lateral to the clavicular head of the sternocleidomastoid muscle. The anode is superior and medial to the cathode (Fig. 4.3).

Control Values
Surface recording, 44 subjects. The upper limit of normal (ULN) is calculated as mean+ 2SD [2] (Fig. 4.5)
   Age (years)
   20–35: latency (ms) 3.2 ± 0.3, ULN 3.8; amplitude (mV) 4.3 ± 3.0.
   36–50: latency (ms) 3.3 ± 0.3, ULN 3.9; amplitude (mV) 3.8 ± 2.4.
   51–65: latency (ms) 3.3 ± 0.3, ULN 3.9; amplitude (mV) 2.7 ± 1.2.
Concentric needle recording, 44 subjects. ULN calculates as mean+ 2SD [2]
   Age (years):
   20–35: latency (ms) 3.6 ± 0.3, ULN 4.2.
   36–50: latency (ms) 3.8 ± 0.3, ULN 4.4.
   51–65: latency (ms) 4.0 ± 0.4, ULN 4.8.

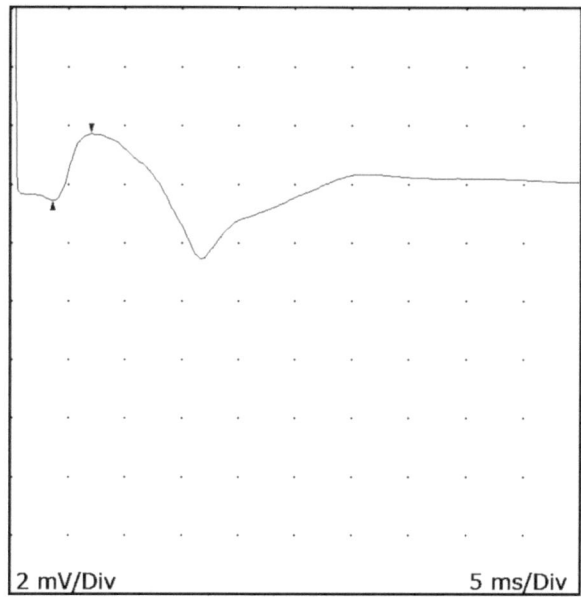

Fig. 4.5 Long thoracic nerve study, control subject, surface recording: DML is 3.8 ms, CMAP amplitude is 2.3 mV

Comments

- Surface recording may be contaminated by volume-conducted CMAPs from other muscles. Measurement of latency by using a concentric needle electrode can be more accurate, but measurement of CMAP amplitude is unreliable for comparisons.
- In concentric needle testing, cover with two fingers of the other hand the nearby intercostal spaces to avoid the risk of pneumothorax or injury to the neurovascular bundle.
- The distance between the stimulating cathode and the recording electrode should be measured with an obstetric caliper.
- Needle electromyography of the serratus anterior muscle is mandatory when long thoracic neuropathy is suspected.
- In long thoracic nerve lesions, the patient has a winged scapula with the vertebral border that stands out especially when trying to push (for instance, against a wall).
- Long thoracic nerve may be involved, even solely, in neuralgic amyotrophy (acute brachial plexitis).

4.3 Suprascapular Nerve Study

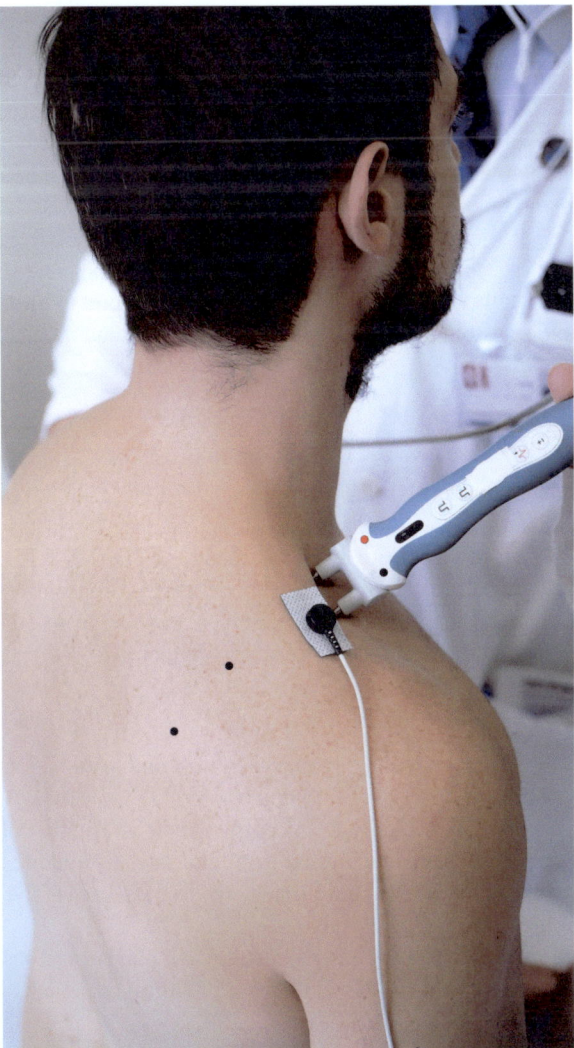

Fig. 4.6 Suprascapular nerve study. The insertion point of the concentric needle electrode in the supraspinatus and infraspinatus are shown

Nerve Fibers Studied and Route
C5 and C6 roots, upper trunk, suprascapular nerve.

Recording Sites
For the supraspinatus muscle, a concentric needle electrode is inserted just above the midpoint of the spine of the scapula until the bone is touched, and then the needle is withdrawn a few millimeters. For the infraspinatus muscle, the needle is inserted 3–4 cm below the midpoint of the scapular spine. G is positioned between the stimulation and recording sites (Fig. 4.6).

Stimulation Site
At Erb's point, the cathode is placed above the clavicle lateral to the clavicular head of the sternocleidomastoid muscle. The anode is superior and medial to the cathode (Fig. 4.6).

Control Values
Subjects 62 [3].
   Supraspinatus. Latency (ms), 2.7 ± 0.5; mean ± 2 SD, 1.7–3.7; range, 1.9–3.8.
   Subjects 60 [3]
   Infraspinatus. Latency (ms), 3.3 ± 0.5 ms; mean ± 2 SD, 2.3–4.3; range, 2.4–4.4.
   The distance between the stimulating cathode and the concentric needle ranges approximately from 7.4 to 12 cm for the supraspinatus and from 10.6 to 15 cm for the infraspinatus when measured by an obstetric caliper [4].

Comments
- The study is performed with the subject sitting and with the arm relaxed by his side.
- The motor response recorded by a concentric needle is usually polyphasic.
- The suprascapular nerve can be injured by trauma such as fracture of the scapula, dislocation of the shoulder, and entrapment syndromes, both at the suprascapular notch and at the spinoglenoid notch.
- The suprascapular nerve injury is seen in volleyball players and baseball pitchers.
- The suprascapular nerve is involved in about one-third of patients with neuralgic amyotrophy (acute brachial plexitis).

4.4 Axillary Nerve Study

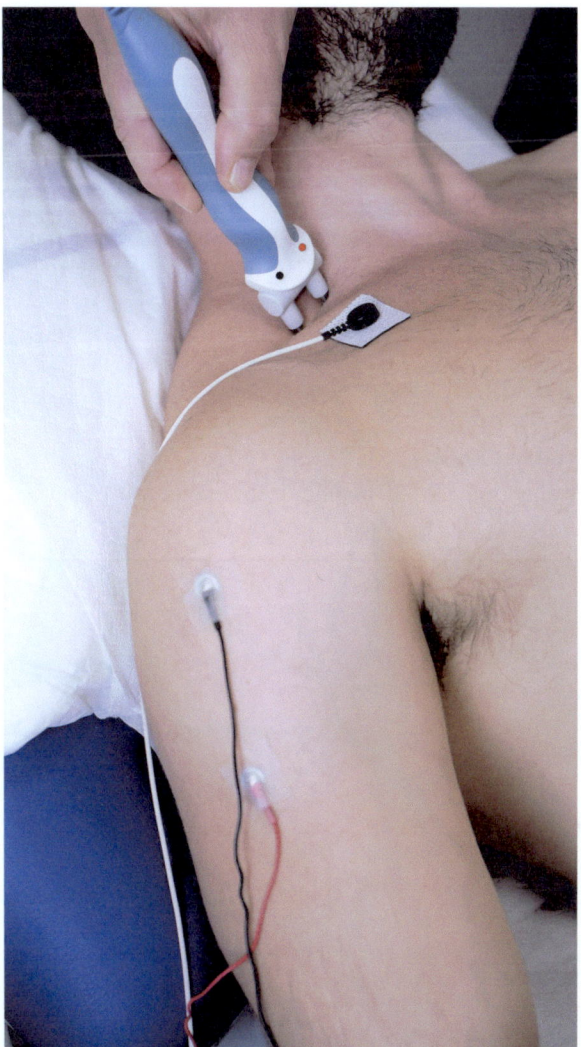

Fig. 4.7 Axillary nerve study

Nerve Fibers Studied and Route
C4 and C5 roots, upper trunk, posterior cord, axillary nerve.

Recording Site
A is placed over the most prominent portion of the middle deltoid muscle; R is placed over the distal tendon insertion of the deltoid. G is placed between the stimulating and recording sites (Fig. 4.7).

Stimulation Site

At Erb's point, the cathode is placed slightly above the clavicle, lateral to the clavicular head of the sternocleidomastoid muscle. The anode is superior and posterior to the cathode (Fig. 4.7). The distance between the stimulating cathode and A ranges from 14.8 to 21 cm measured by an obstetric caliper with the arm by the side [5].

Control Values

62 subjects [6].

Latency (ms), 3.9 ± 0.5; mean ± 2 SD, 2.9–5.0; range, 2.9–5.2 (Fig. 4.8).
154 subjects (reference electrode over the acromion). ULN (95th percentile), lower limit of normal (LLN) (5th percentile) [7].

Males:
Latency (ms): ULN 5.
Amplitude (mV): LLN 7.
Females:
Latency (ms): ULN 3.5.
Amplitude (mV): LLN 6.5.
Side-to-side abnormal CMAP amplitude difference: >40%.

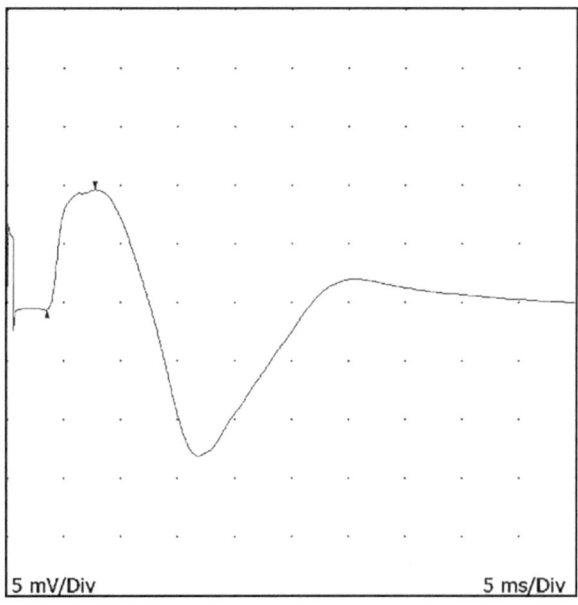

Fig. 4.8 Axillary nerve study, control subject: DML is 3.5 ms, CMAP amplitude is 10.3 mV

Comments

- The best position to examine the subject is supine with the elbow extended and the forearm supinated.
- It may be difficult to achieve a supramaximal stimulation at Erb's point.

- The axillary nerve may be involved in shoulder dislocation or fracture of the humeral neck.
- Axillary neuropathy occasionally follows general anesthesia or sleeping in a prone position with the arm raised above the head.
- The axillary nerve is commonly involved in neuralgic amyotrophy (acute brachial plexitis).

4.5 Musculocutaneous Nerve Study

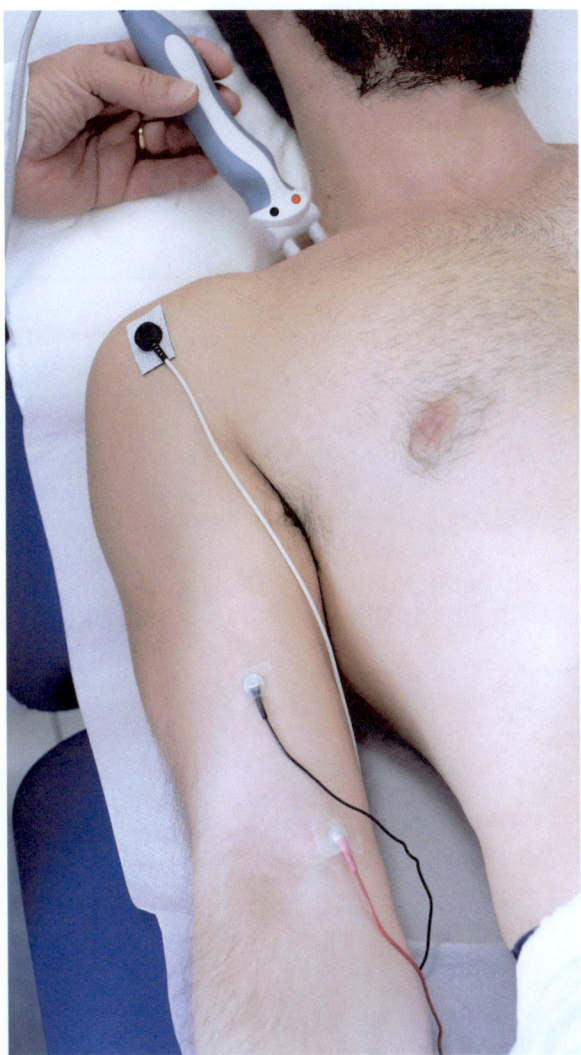

Fig. 4.9 Musculocutaneous nerve study

Nerve Fibers Tested and Route
C5 and C6 roots, upper trunk, lateral cord, musculocutaneous nerve.

Recording Site
A is placed just distal to the midportion of the biceps muscle; R is placed distally in the antecubital fossa on the biceps tendon. G is placed between the recording and stimulating electrodes (Fig. 4.9).

Stimulation Site
At Erb's point lateral to the clavicular head of the sternocleidomastoid muscle in the supraclavicular fossa. The anode is superomedial (Fig. 4.9).

The distance between stimulating electrodes and A ranges approximately from 23.5 to 29.0 cm when measured with an obstetrical caliper and the arm by the side [8].

Control Values
62 subjects [9].

Latency (ms), 4.5 ± 0.6; mean ± 2 SD, 3.3–5.7; range, 3.2–5.9 (Fig. 4.10).

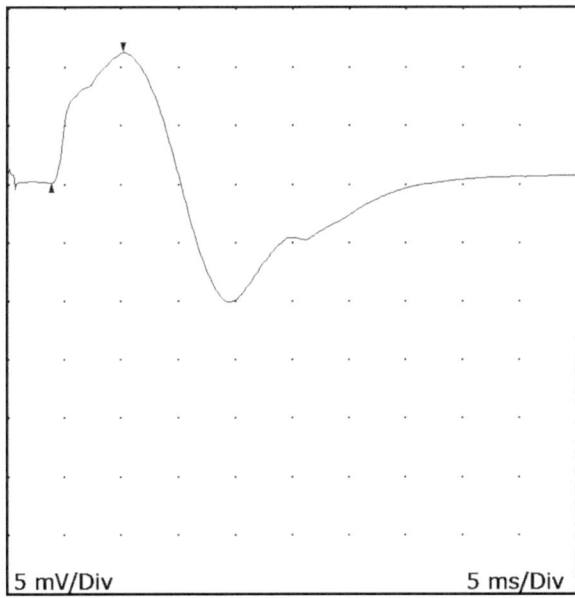

Fig. 4.10 Musculocutaneous nerve motor study. Control subject: DML is 3.8 ms, CMAP amplitude is 11.1 mV

Comments
- It may be difficult to achieve a supramaximal stimulation at Erb's point.
- Musculocutaneous neuropathy may complicate shoulder dislocation and is sometimes associated with axillary nerve involvement.

- Isolated musculocutaneous neuropathies are uncommon.
- The study is useful because it provides information concerning the C5–C6 root. Moreover, traction injuries of the brachial plexus give information about the integrity of the upper trunk and lateral cord.

4.6 Lateral Cutaneous Nerve of the Forearm Study

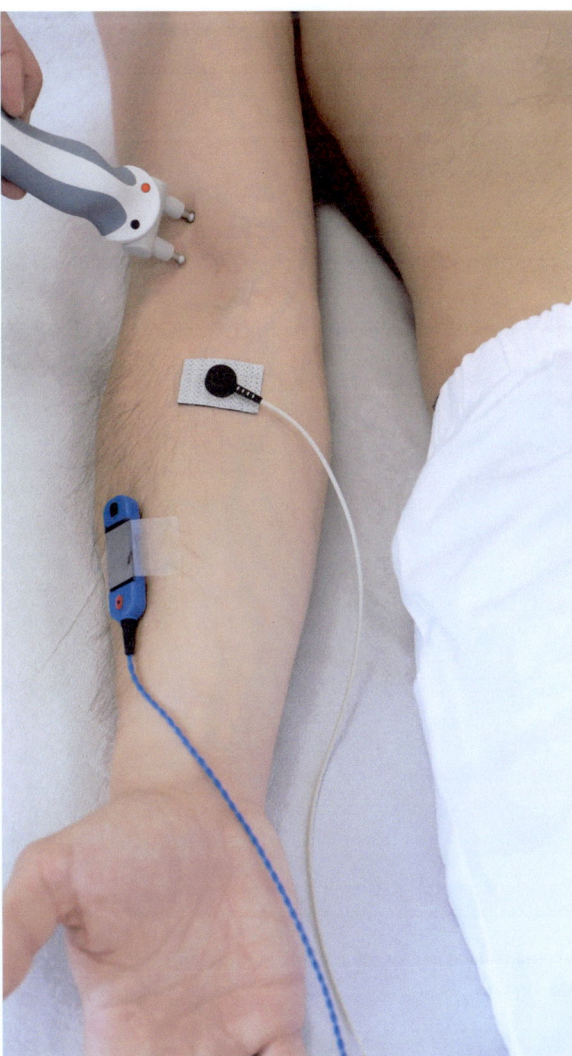

Fig. 4.11 Lateral cutaneous nerve of the forearm study

Nerve Fibers Tested and Route

Lateral cutaneous nerve of the forearm, musculocutaneous nerve, lateral cord, upper trunk, C5 and C6 roots.

Recording Site

A 3 cm electrode bar is placed 12 cm distal to the stimulating cathode along a straight line to the radial artery at the wrist; G is placed between the recording electrode and the stimulating sites (Fig. 4.11).

Stimulation Site

At the antecubital fossa, just lateral to the biceps tendon (Fig. 4.11).

Control Values

60 nerves of 30 normal adults [10].

Latency (ms), 1.8 ± 0.1; mean +2SD, 2.0; range, 1.6–2.1.
Side-to-side difference (mean +2SD): <0.3 ms.
Peak-to-peak amplitude (µV), 24 ± 7.2; mean -2SD, 9.6; range, 12–50 (Fig. 4.12).

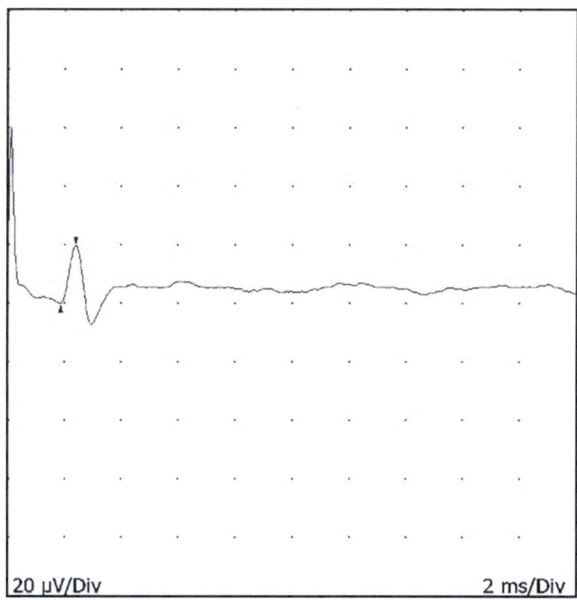

Fig. 4.12 Lateral cutaneous nerve of the forearm study, control subject: SNAP latency is 1.9 ms, SNAP amplitude is 16.3 µV

Comments

- The lateral antebrachial cutaneous nerve is the terminal sensory branch of the musculocutaneous nerve, and it supplies the skin of the lateral and ventral radial border of the forearm stopping at the wrist.

- To avoid a motor artifact, it is better to keep the stimulator close to the biceps tendon; in this way, the contraction of the brachioradialis muscle can be avoided.
- Side-to-side comparisons of the SNAP amplitude and latency are helpful to assess abnormalities.
- The nerve may be compressed by carrying a heavy bag with the strap across the elbow crease or damaged by a venipuncture of the cephalic vein.
- It is involved also in brachial plexus lesions involving the upper trunk or lateral cord.

4.7 Radial Nerve Motor Study

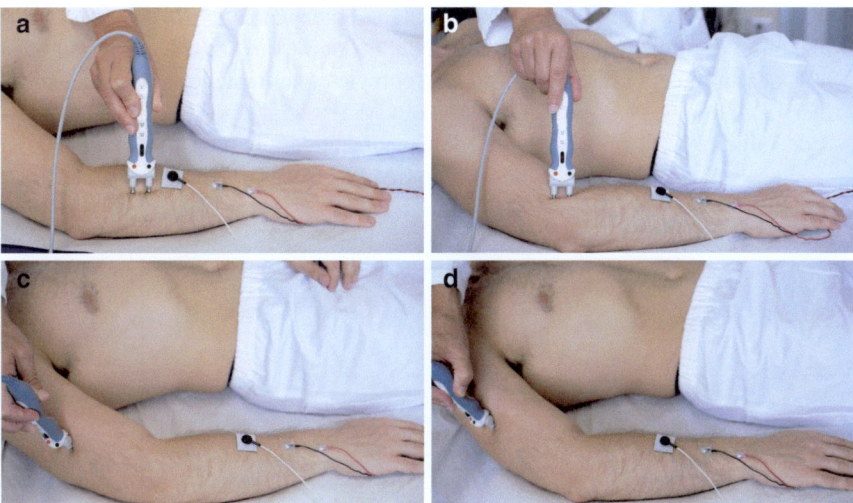

Fig. 4.13 Radial nerve motor study. (**a**) Stimulation at the forearm, (**b**) stimulation at the elbow, (**c**) stimulation below the spiral groove, (**d**) stimulation above the spiral groove

Nerve Fibers Tested and Route
C7 and C8 roots, middle and lower trunks, posterior cord, radial nerve, posterior interosseous branch.

Recording Site
With hand pronated, A is placed over the extensor indicis proprius muscle about 4 cm proximal to the ulnar styloid process; R is placed over the ulnar styloid. G is placed between the stimulating and recording sites (Fig. 4.13).

Stimulation Sites
Forearm, over the ulna, 4–6 cm proximal to the A; elbow, in the groove between the biceps and brachioradialis muscles; below spiral groove, lateral midarm, between the biceps and triceps muscles; above spiral groove, posterior proximal arm over the humerus (Fig. 4.13).

Control Values [11]
Distal latency (ms): ≤2.9.
  Distal CMAP amplitude (mV): ≥2.0.
  CV, all segments (m/s): ≥49.

Comments

- Concentric needle recording has also been described. In this case, it is important to ensure that changes in the CMAP waveform and amplitude from different stimulation sites are not due to needle movement.
- The surface-recorded CMAPs may show an initial positive deflection caused by a volume-conducted CMAP generated by adjacent muscles innervated by the stimulated radial nerve (Fig. 4.14). It is debated where to measure the onset latency of a CMAP with a positive onset, and independently if it is measured at the onset or at the peak of the positive deflection, it should be kept in mind that the measurement of distal latency is not completely reliable for assessing a distal nerve pathology. However, in the case of CMAPs with a positive onset, conduction velocity (CV) can be calculated by paying attention to the fact that the onset latency is measured at the same point in CMAPs from distal and proximal stimulation sites (Fig. 4.14).

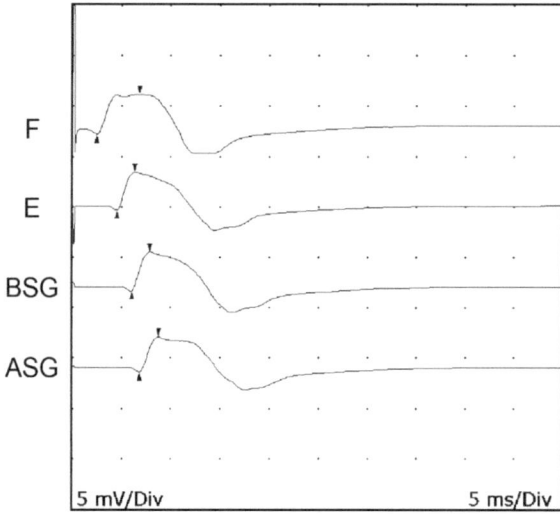

Fig. 4.14 Radial nerve motor study, control subject. Stimulation at forearm (F), elbow (E) below spiral groove (BSG), above spiral groove (ASG). DML is 2.5 ms, distal CMAP amplitude is 4 mV, CV in nerve segments is between 55 and 62 m/s

- Surface-measured distances often are inaccurate, especially at proximal stimulation sites. Distances to the sites below and above the spiral groove should be measured with an obstetric caliper.

- Damage to the radial nerve may involve the main trunk of the nerve in the axilla or upper arm or the two major terminal branches: the posterior interosseous and the superficial radial nerve.
- The radial nerve may be compressed at the spiral groove as in "Saturday night palsy" (in this case, the triceps muscles are usually spared) or injured by fractures of the humerus.
- The posterior interosseous may be compressed by soft tissue masses (lipoma, enlarged bursa, ganglia arising from the elbow joint), at the Frohse arcade during prolonged pronation-supination movements.

4.8 Posterior Cutaneous Nerve of the Forearm Study

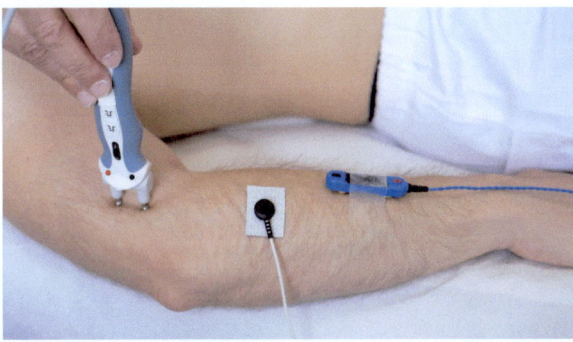

Fig. 4.15 Posterior cutaneous nerve of the forearm study

Nerve Fibers Studied and Route
Posterior cutaneous nerve of the forearm; radial nerve; posterior cord; upper, middle, and lower trunks; and C5, C6, C7, and C8 roots.

Recording Site
With the forearm pronated, A is placed 12 cm below the stimulating cathode at midway between the ulnar and radial styloid processes along a line to the dorsum of the wrist; R is placed 3 cm distally. G is placed between the recording electrodes and the stimulating sites (Fig. 4.15).

Stimulation Site
At the elbow about 2 cm above the lateral epicondyle between the medial border of the lateral head of the triceps muscle and the biceps muscle (Fig. 4.15).

Control Values

22 subjects [12].

Latency (ms), 1.9 ± 0.3; mean +2SD, 2.5; range, 1.5–2.4.
Peak-to-peak amplitude (μV), 8.6 ± 3.9; range, 5–20 μV (Fig. 4.16).

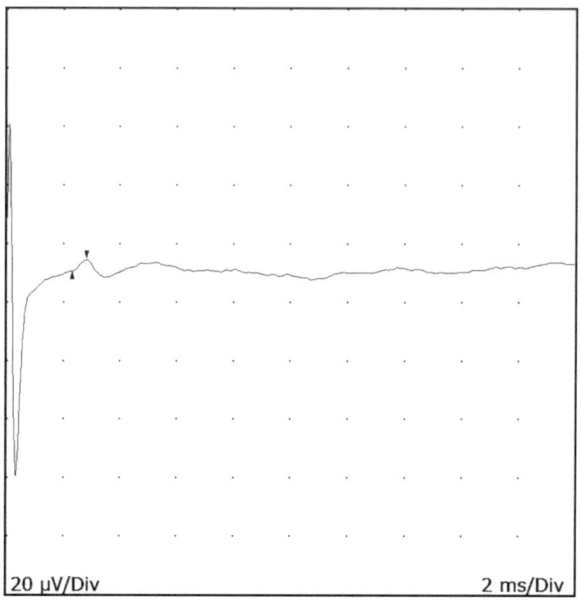

Fig. 4.16 Posterior cutaneous nerve of the forearm study, control subject: SNAP latency is 2.3 ms, SNAP amplitude is 5.0 μV

Comments

- The nerve originates from the radial nerve at the lower third of the arm and becomes superficial between the lateral epicondyle of the humerus and olecranon; then it runs on the posterior dorsolateral surface of the forearm toward the midline of the wrist.
- The SNAP may be difficult to record because of motor artifacts due to excitation of nearby muscles. Lower stimulus intensity and repositioning of recording and stimulating electrodes may be necessary.
- The exploration of this nerve can have a role in the localization of the site of radial nerve injury in addition to radial motor conduction study, superficial radial sensory conduction study, and needle electromyography.

4.9 Radial Nerve Sensory Study

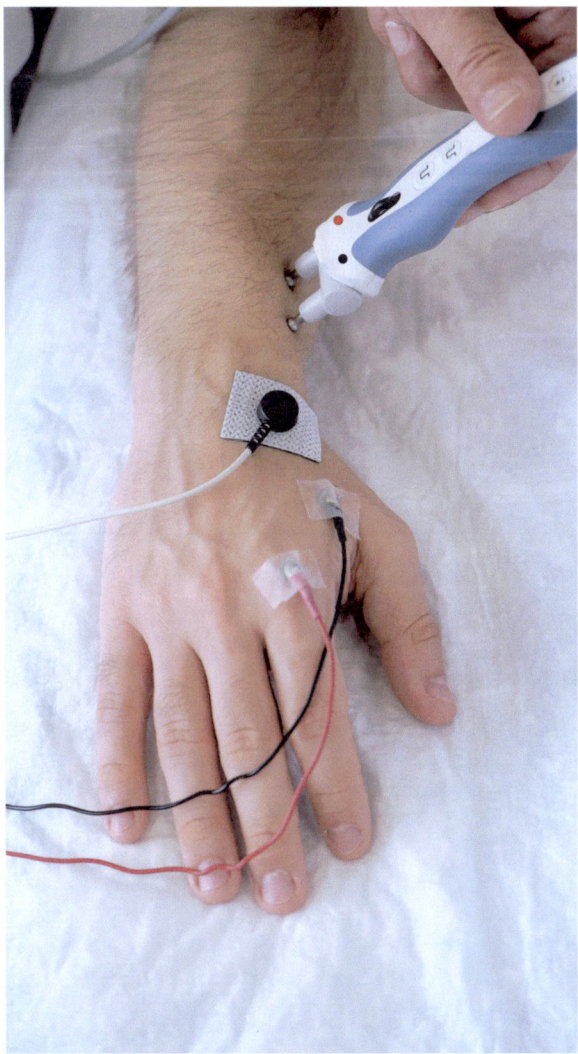

Fig. 4.17 Radial nerve sensory study

Nerve Fibers Tested and Route
Superficial radial nerve, radial nerve, posterior cord, upper trunk, C6 root.

Recording Site
A is placed over the superficial radial nerve that can be felt as it runs over the tendon of the extensor pollicis longus at the wrist; R is placed 3–4 cm distally slightly proximal to the head of the 2nd metacarpal bone. G is placed on the dorsum of the hand between the stimulating and recording electrodes (Fig. 4.17).

Stimulation Site
The cathode is placed over the radius 10 cm proximal to A. The anode is proximal (Fig. 4.17).

Control Values
49 subjects [13].
  Latency (ms), 1.8 ± 0.3; mean + 2SD, 2.4.
  Peak-to-peak amplitude (µV), 31 ± 20 µV; range, 13–60 (Fig. 4.18).

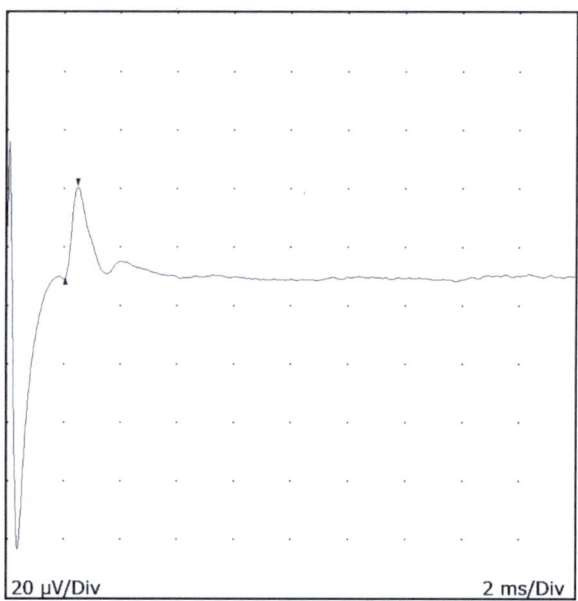

Fig. 4.18 Radial nerve sensory study, control subject: SNAP latency is 2.1 ms, SNAP amplitude is 32.4 µV

Comments
- The nerve can be felt and localized by asking the subject to extend the thumb.
- The superficial radial nerve is usually damaged at the wrist by compression of wristwatch bands, handcuffs, bracelets, and tight plaster casts.

- An injury to the nerve may occur during deQuervain's tenosynovectomy.
- May be spared in lesions at the spiral groove because the superficial branch may take off more proximally or because of differential involvement of fascicles within the radial nerve.
- May be involved in lesions of the posterior cord and upper trunk of the brachial plexus.
- It is spared in posterior interosseous neuropathy.

4.10 Medial Cutaneous Nerve of the Forearm Study

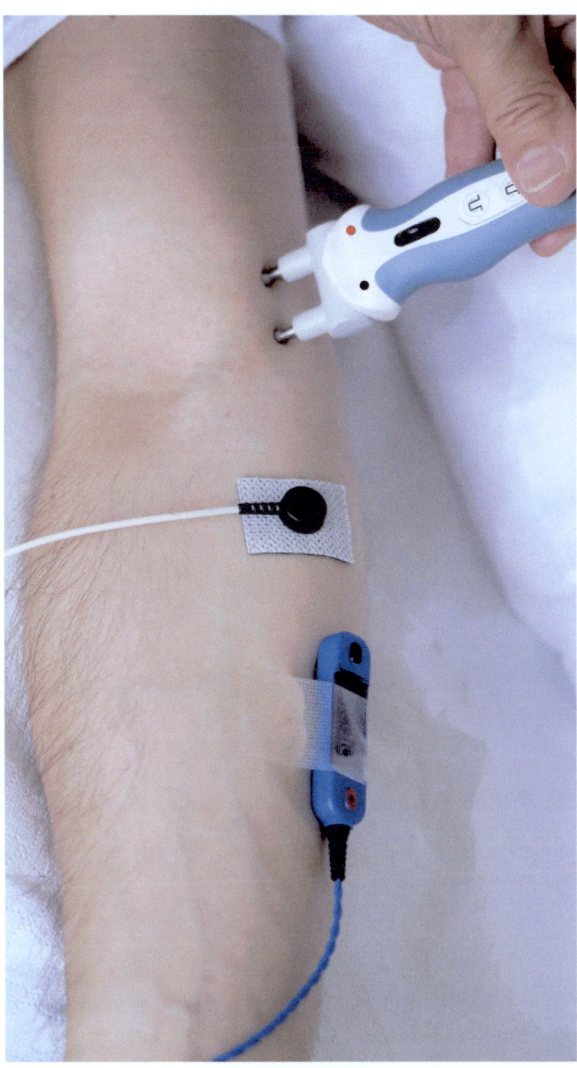

Fig. 4.19 Medial cutaneous nerve of the forearm sensory study

Nerve Fibers Tested and Route
Medial cutaneous nerve of the forearm, medial cord, lower trunk, C8 and T1 roots.

Recording Site
A 3 cm bar electrode is placed on the anteromedial surface of the forearm with A 14 cm below the stimulation site; R is placed distally. G is placed between the recording electrode and the stimulating electrodes (Fig. 4.19).

Stimulation Site
The cathode is placed medial to the biceps tendon, and the anode is proximal (Fig. 4.19).

Control Values
40 subjects [14].

Latency (ms), 1.54 ± 0.17; mean + 2SD, 1.88; range, 1.2–2.0.

Peak-to-peak amplitude (μV), 18.8 ± 7.1; mean −2SD, 4.6; range, 13–60 (Fig. 4.20).

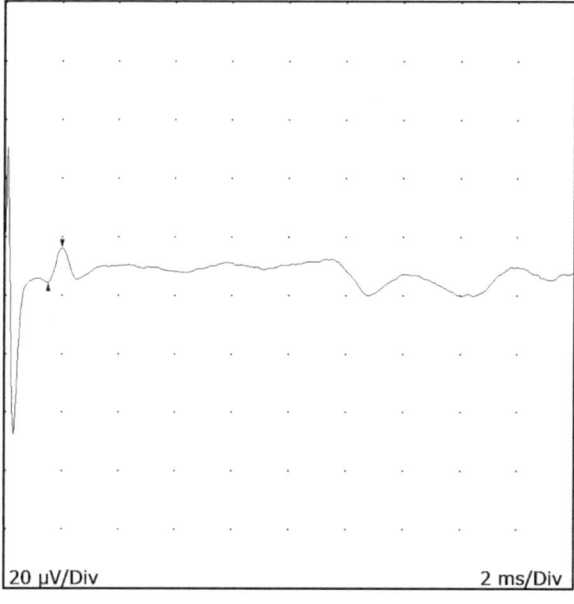

Fig. 4.20 Medial cutaneous nerve of the forearm study, control subject: the SNAP latency is 1.5 ms. The SNAP amplitude is 13.2 μV

Comments

- The SNAP of the medial cutaneous nerve of the forearm has usually a lower amplitude than the SNAP of the lateral cutaneous nerve of the forearm.

- Side-to-side comparisons of the SNAP amplitude and latency are helpful to assess abnormalities.
- The medial cutaneous nerve of the forearm can be damaged during infusion therapy using the basilic vein.
- The study may be abnormal in lesions of the medial cord or lower trunk of the brachial plexus.
- The SNAP is absent or with reduced amplitude in neurogenic thoracic outlet syndrome.

4.11 Ulnar Nerve Motor Study

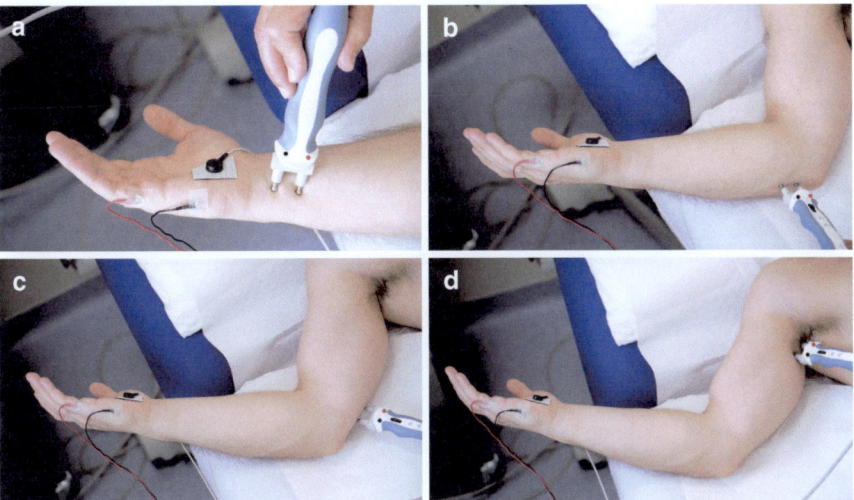

Fig. 4.21 Ulnar nerve motor study. (**a**) stimulation at wrist, (**b**) stimulation below elbow, (**c**) stimulation above elbow, (**d**) stimulation at axilla

Nerve Fibers Tested and Route
C8 and T1 roots, lower trunk, medial cord, ulnar nerve.

Recording Site
Abductor digiti minimi (ADM) muscle. A is placed over the hypothenar eminence, halfway between the level of the pisiform bone and the fifth metacarpophalangeal joint; R is placed slightly distal to the fifth metacarpophalangeal joint. G is placed on the dorsum of the hand. If the stimulus artifact interferes with the recording, G can be placed between A and the stimulating cathode (Fig. 4.21).

Stimulation Sites

Wrist, cathode at medial wrist, adjacent to the flexor carpi ulnaris tendon 8 cm proximal to A; below elbow, cathode 3–4 cm distal to the medial epicondyle; above elbow, cathode over the medial humerus, between the biceps and triceps muscles, at a distance of 10–12 cm from below elbow measured in a curve behind the medial epicondyle; axilla, cathode in the medial axilla approximately 10 cm from above elbow. The anode is proximal at all stimulation sites.

The optimal position for the study and distance measurement is with the elbow flexed between 70° and 90° (Fig. 4.21).

Control Values

248 subjects [15].

Distal motor latency (DML) (ms), 3.0 ± 0.3; mean + 2SD, 3.6; 97th percentile 3.7.

Distal amplitude (mV), 11.6 ± 2.1; mean −2SD, 7.4; 3rd percentile 7.9 mV.

CV (m/s):

mean -2SD, 51; 3rd percentile 52.

AE-BE 61 ± 9; mean -2SD, 43; 3rd percentile 43.

Ax-BE 61 ± 7; mean -2SD, 47; 3rd percentile 50 (Fig. 4.22).

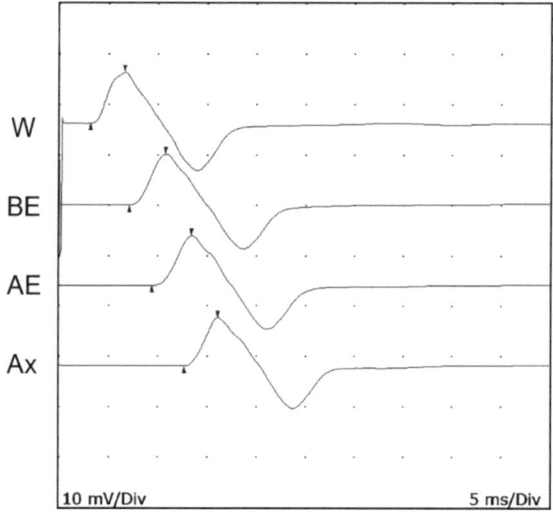

Fig. 4.22 Ulnar motor nerve study, control subject with stimulation at wrist (W), below elbow (BE), above elbow (AE), and axilla (Ax): DML is 3.1 ms, distal CMAP amplitude is 10.1 mV, CV in the BE-W segment is 59 m/s, in the AE-BE segment is 54 m/s, in the Ax-AE segment is 56 m/s

Comments

- Another proximal stimulation site of the ulnar nerve is at Erb's point. At this site, supramaximal stimulation cannot always be reached in thickset subjects. When calculating conduction velocity in Erb's point to the axilla segment, the distance should be measured by an obstetric caliper.
- If the study is performed in a straight-elbow position, a factitious conduction slowing across the elbow can be calculated due to the underestimation of the true nerve length.
- Stimulation at the below-elbow site must be at least 3 cm distal to the medial epicondyle to be sure that stimulation is distal to the cubital tunnel, a common site of ulnar nerve compression at the elbow.
- Higher current intensity usually is needed to reach supramaximal stimulation at the below-elbow site compared with the wrist, especially in obese or muscular subjects because, at this location, the nerve is deep to the flexor carpi ulnaris muscle. If a comparable CMAP amplitude cannot be obtained, a conduction block can be suspected. Conduction block can be excluded if the CMAP amplitude from above-elbow stimulation is comparable with the CMAP amplitude from wrist stimulation.
- Ulnar neuropathy at the elbow may be due to compression at three sites: the medial intermuscular septum, epicondylar groove, and cubital tunnel.
- A short nerve segment incremental stimulation (inching technique) can be performed to localize the site of ulnar neuropathy at the elbow. First, the nerve's course is mapped out with the elbow fixed at 90° by submaximal stimulation at sequential sites, and the nerve course is traced with a pen. Then, supramaximal stimulation is performed in 1–2 cm increments along the nerve taking care not to use excessive stimulation. The upper limit of normal segmental latency change for a 1 cm increment is 0.4 ms [16], and for a 2 cm increment, it is 0.63 ms. [17] Abrupt changes in waveform and amplitude may indicate a focal conduction block.
- If the CMAP amplitude from the below-elbow and above-elbow stimulation sites is smaller than that at the wrist, consider a Martin-Gruber anastomosis.
- The conduction to the abductor digiti minimi is normal in Guyon's channel with entrapment of the deep motor branch of the ulnar nerve. In this case, the conduction to the first dorsal interosseous should be studied.

4.12 Deep Ulnar Motor Branch Study

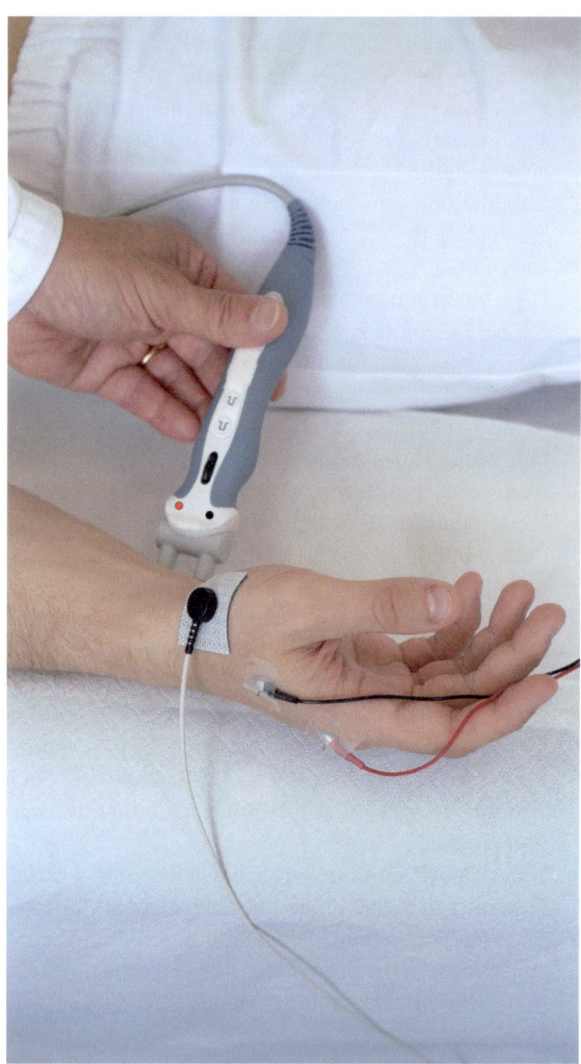

Fig. 4.23 Deep ulnar motor branch study

Nerve Fibers Tested and Route
C8 and T1 roots, lower trunk, medial cord, ulnar nerve, deep ulnar motor branch.

Recording Site
First dorsal interosseous muscle (dorsal web space between the thumb and index finger): A placed over the muscle belly and R placed over the second metacarpophalangeal joint of the thumb. G is placed between A and the stimulating cathode (Fig. 4.23).

4.12 Deep Ulnar Motor Branch Study

Stimulation Site

The cathode over the ulnar nerve at the proximal wrist crease. The anode is proximal (Fig. 4.23).

Control Values

188 subjects, 373 nerves [18].
  Latency (ms): ULN 4.5.
  Amplitude (mV): LLN 6.

The ULN of the latency difference of CMAP from the first dorsal interosseous and CMAP from the abductor digiti minimi with the same stimulation point is 2.0 ms (mean 0.9 ms, range 0.2–2).

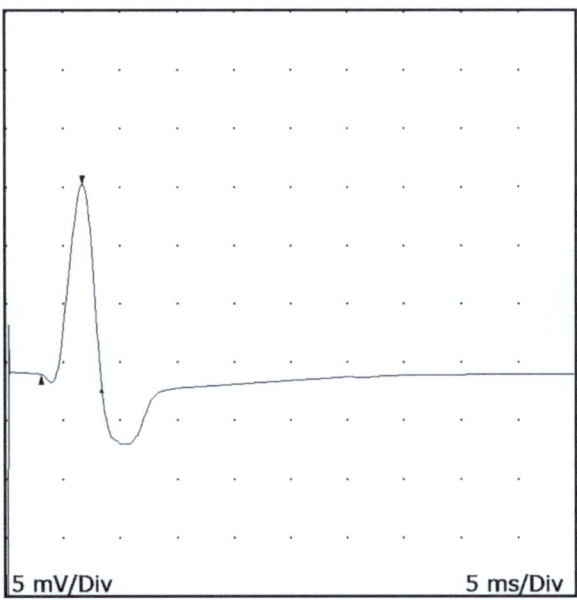

Fig. 4.24 Deep ulnar motor branch study, control subject: DML is 3.1 ms, CMAP amplitude is 16.2 mV

Comments

- The CMAP recorded from the first dorsal interosseous may present an initial positive deflection caused by a volume-conducted CMAP generated by adjacent muscles innervated by the stimulated ulnar nerve (Fig. 4.24). It is debated where to measure the onset latency of a CMAP with a positive onset; however, independently of whether it is measured at the initial deflection from the baseline (as in 4.24) or at the peak of the positive deflection, it should be kept in mind that in this case, the measurement of distal latency may not be completely reliable for calcu-

lating the latency difference of CMAP from the first dorsal interosseous and CMAP from the abductor digiti minimi to assess a distal nerve pathology.
- CMAP amplitude recorded from the first dorsal interosseous is usually greater than that recorded from the abductor digiti minimi. If the CMAP amplitude at the below-elbow stimulation is smaller than that at the wrist, consider a Martin-Gruber anastomosis. Anomalous innervation of the first dorsal interosseous by the radial nerve through the radial superficial sensory branch is sometimes present. The deep ulnar motor branch often is preferentially affected in lesions of the ulnar nerve at Guyon's canal [19, 20]. Because nerve fascicles can be differentially involved in ulnar neuropathies at the elbow, recording from the FDI may be more useful than recording from the ADM for demonstrating focal slowing of the ulnar nerve across the elbow [21, 22].

4.13 Ulnar Nerve Sensory Study to Digit 5

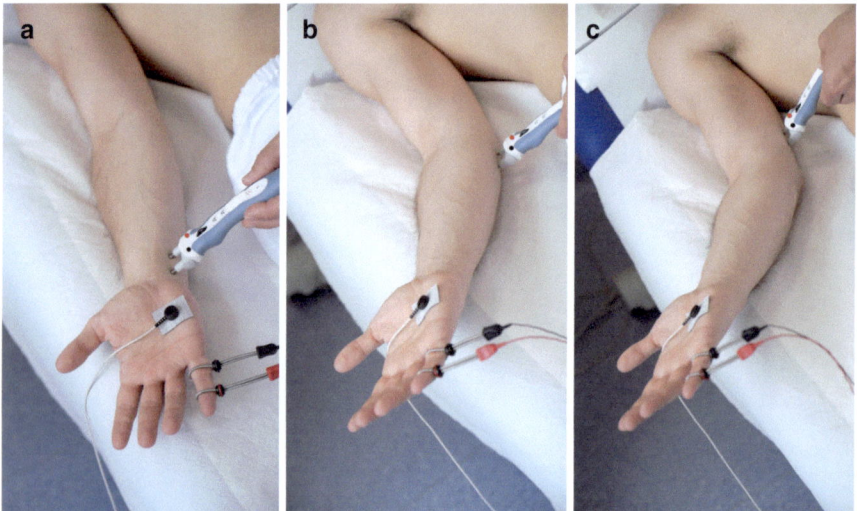

Fig. 4.25 Ulnar nerve sensory study to digit 5. (**a**) Stimulation at the wrist, (**b**) stimulation below the elbow, (**c**) stimulation above the elbow

Nerve Fibers Tested and Route
Digital nerves, ulnar nerve, medial cord, lower trunk, C8.

Recording Site
By using ring electrodes with A placed slightly distal to the metacarpophalangeal joint of the fifth digit and R placed 3–4 cm distally over the distal interphalangeal joint. G is placed on the palm or dorsum of the hand (Fig. 4.25).

4.13 Ulnar Nerve Sensory Study to Digit 5

Stimulation Sites

Medial wrist (W), adjacent to the flexor carpi ulnaris tendon 3 cm above the distal crease of the wrist. Stimulation can be performed also at the below- and above-elbow sites, similar to the ulnar motor study (Fig. 4.25).

Control Values

130 nerves of 65 subjects [23].
Latency (ms), 2.54 ± 0.29; mean + 2SD, 3.1.
Baseline to negative peak amplitude (μV), 35.0 ± 14.7; mean −2SD, 18.
CV (m/s), 54.8 ± 5.3; mean −2SD, 44 (Fig. 4.26).
Stimulation at the below- and above-elbow sites.
20 subjects [24].
Mean + 2SD value for latency increment for 10 cm segment across the elbow: 1.8 ms.
LLN for CV BE-W segment: 59 m/s,
LLN for CV AE-BE segment: 50 m/s.
ULN of the SNAP amplitude decrement for the BE-W segment: 74% (range 12–72%).
ULN of the SNAP amplitude decrement for the AE-BE segment: 41% (range 0–50%) (Fig. 4.27).

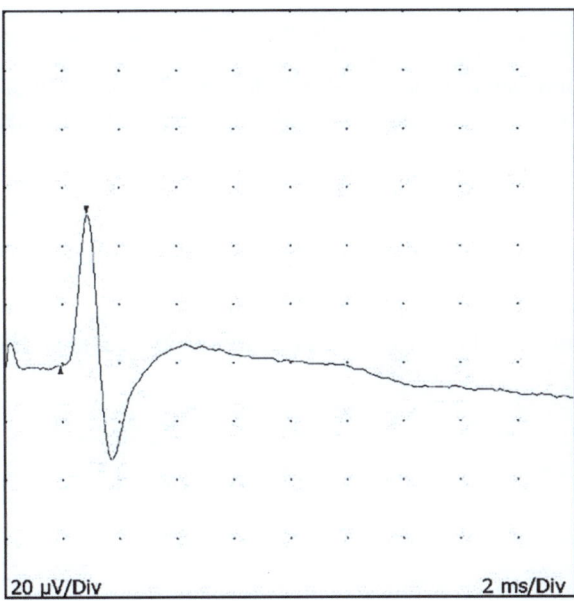

Fig. 4.26 Ulnar nerve sensory study to digit 5 with stimulation at wrist, control subject: SNAP latency is 1.9 ms, SNAP amplitude is 51 μV

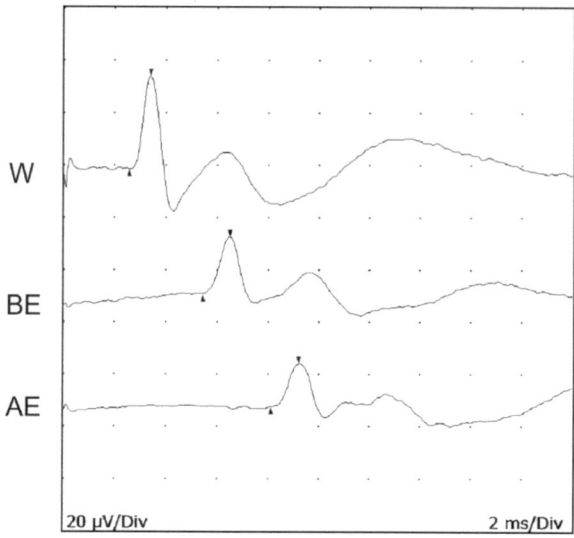

Fig. 4.27 Ulnar nerve sensory study to digit 5 with stimulation at wrist (W), below elbow (BE) and above elbow (AE), control subject: SNAP amplitude is 37.3 μV from W stimulation, 24.2 μV from BE stimulation, and 17.1 μV from AE stimulation. CV in the W-D5 segment is 54 m/s, in the BE-W segment is 62 m/s, and in the AE-BE segment is 60 m/s

Comments

- The antidromic study is described. For the orthodromic study, stimulation and recording sites are reversed. Orthodromic sensory potentials have lower amplitudes.
- Physiological temporal dispersion has more phase cancellation effect in sensory than in motor conduction studies causing a significant decrease in the SNAP amplitude with increasing distance from the stimulation and recording sites (Fig. 4.27).
- This study may be abnormal in ulnar neuropathy or thoracic outlet syndrome.
- This study may be abnormal, in the presence of a normal ulnar DML, in lesions at the exit of Guyon's channel.

4.14 Ulnar Dorsal Cutaneous Nerve Study

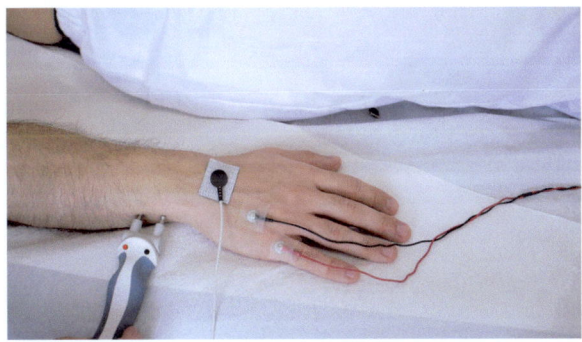

Fig. 4.28 Ulnar dorsal cutaneous nerve study

Nerve Fibers Tested and Route
Dorsal cutaneous branch, ulnar nerve, medial cord, lower trunk, C8.

Recording Site
A is placed on the dorsum of the hand over the web space between the little and ring finger. R is placed 3 cm distally at the base of the little finger. G is placed on the dorsum of the hand between the stimulating and recording electrodes (Fig. 4.28).

Stimulation Site
With the hand pronated, the cathode is positioned slightly proximal and inferior to the ulnar styloid 8–10 cm from A. The anode is proximal (Fig. 4.28).

Control Values
30 subjects [25].
 Latency (ms), 2.0 ± 0.3; mean + 2 SD, 2.6.
 Amplitude (µV), 29 ± 6 µV; mean −2 SD, 17 (Fig. 4.29).

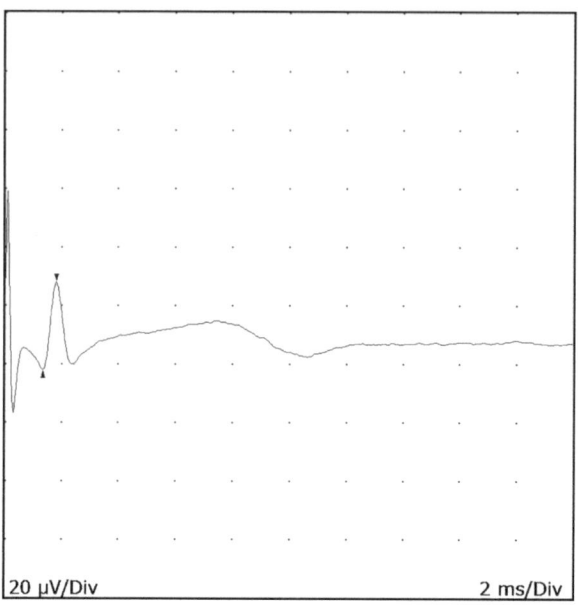

Fig. 4.29 Ulnar dorsal cutaneous nerve study, control subject: SNAP amplitude is 31.4 µV, SNAP latency is 1.4 ms

Comments

- The nerve is superficial, and supramaximal stimulation can be reached with low stimulus intensities.
- The SNAP may be obscured by a motor artifact; in this case, decreasing the stimulus intensity can be useful.
- The ulnar dorsal cutaneous branch arises about 5 cm above the wrist, and it is always spared in lesions of the ulnar nerve at Guyon's canal.
- Because nerve fascicles can be differentially involved in ulnar neuropathies at the elbow, the dorsal cutaneous branch is not always involved at this site [26].
- The dorsal cutaneous branch may arise from the superficial radial nerve, and this may be a source of error in the interpretation of conduction studies.

4.15 Anterior Interosseous Nerve Study

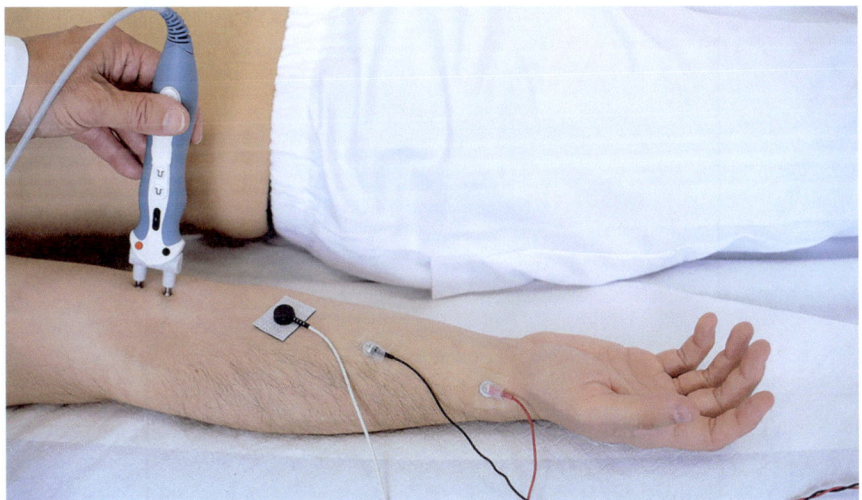

Fig. 4.30 Anterior interosseous nerve study

Nerve Fibers Tested and Route
C7, C8, and T1 roots, middle and lower trunks, medial and lateral cords, median nerve, anterior interosseous nerve.

Recording Site
A is on the lateral forearm on the flexor pollicis longus over about the distal one-third of the distance from the antecubital fossa to the volar crease of the wrist, and R is over the distal tendon of the flexor pollicis longus. G is between the stimulating and recording sites (Fig. 4.30).

Stimulation Site
Medial to the tendon of the biceps at the antecubital fossa (Fig. 4.30).

Control Values
25 subjects [27].
  Latency (ms), 2.6 ± 0.43; range, 1.8–3.6; ULN 4.0.
  Amplitude (mV), 5.6 ± 1.16; range, 3.8–7.5; LLN 2.5 (Fig. 4.31).

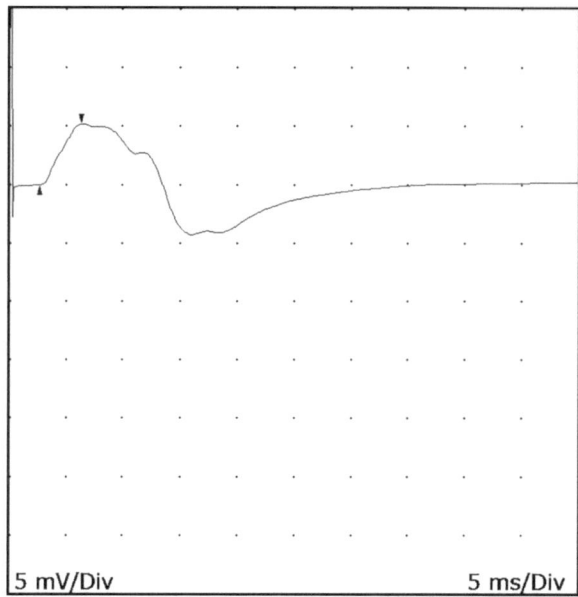

Fig. 4.31 Anterior interosseous nerve study, control subject: DML is 2.7 ms, and CMAP amplitude is 5.3 mV

Comments

- The anterior interosseous is a purely motor nerve, and its damage causes the loss of pinching with the thumb and index finger.
- An initially positive deflection may be recorded because the active electrode can be distant from the muscle end plate region.
- Causes of anterior interosseous neuropathies are traumatic, compressive, and excessive forearm exercise.
- Anterior interosseous neuropathy, even isolated, has been described in neuralgic amyotrophy (acute brachial plexitis).

4.16 Median Nerve Motor Study

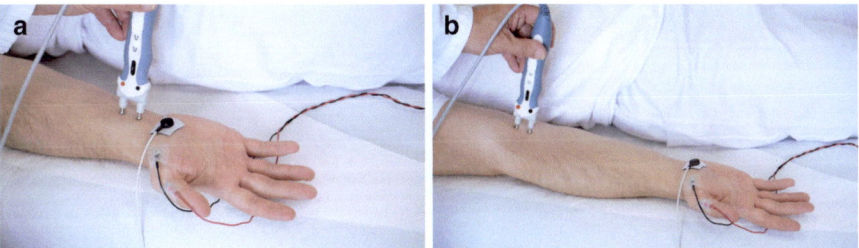

Fig. 4.32 Median nerve motor study. (**a**) Stimulation at the wrist, (**b**) stimulation at the antecubital fossa

Nerve Fibers Tested and Route
C8 and T1, lower trunk, medial cord, median nerve.

Recording Site
A is placed over the abductor pollicis brevis (APB) muscle belly halfway between the metacarpophalangeal joint of the thumb and the distal wrist crease; R is placed over the first metacarpophalangeal joint. G is usually placed on the dorsum of the hand. If the stimulus artifact interferes with the recording, G can be placed between A and the stimulating cathode (Fig. 4.32).

Stimulation Sites
At the wrist between the tendons of the flexor carpi radialis and palmaris longus 8 cm from A; at the antecubital fossa just medial to the brachial artery pulse. Other proximal sites of stimulation are at the axilla and Erb's point (Fig. 4.32).

Control Values
249 subjects [28].
  Latency (ms), 3.7 ± 0.5; mean + 2SD, 4.7; 97th percentile 4.5.
  Amplitude (mV), 10.2 ± 3.6; mean −2SD, 3.0; 3rd percentile 4.1.
  CV (m/s), 57 ± 5; mean −2SD, 47; 3rd percentile 49 (Fig. 4.33).

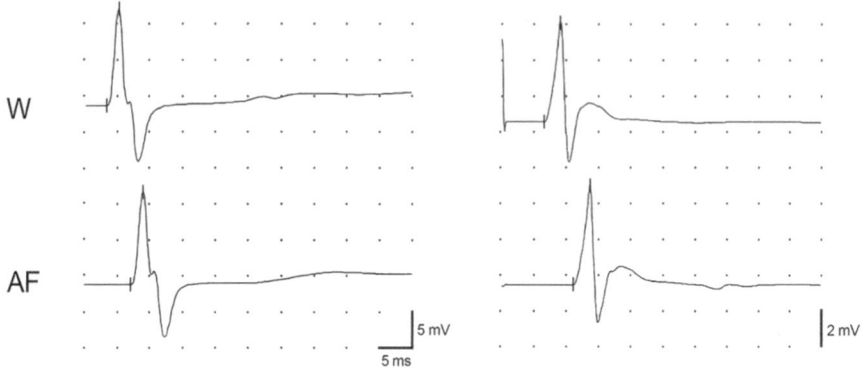

Fig. 4.33 Median nerve motor study with stimulation at the wrist (W) and antecubital fossa (AF). On the left are tracings recorded from a control subject: DML is 3.8 ms, distal CMAP amplitude is 13 mV, and CV is 58 m/s. On the right are tracings from a CTS patient: DML is increased to 6.6 ms, distal CMAP amplitude is 5.8 mV, and CV is borderline/reduced (48 m/s)

Comments

- In median motor nerve conduction, the CMAP waveform should be similar after distal and proximal stimulation. Excessive stimulation at the wrist and antecubital fossa may result in co-stimulation of the ulnar nerve suggested by an initial positive onset. In this case, reduce the stimulation intensity. It is more difficult to avoid co-stimulation of the ulnar nerve at the axilla and Erb's point.
- Median neuropathies occur at the axilla, upper arm, the elbow region, and at the carpal tunnel that is by far the commonest.
- Median neuropathy at the elbow is rarely due to a supracondylar spur and ligament of Struthers or when the median nerve passes between the two heads of the pronator teres muscle (pronator teres syndrome) that is quite a controversial entity.
- In a review study, the evaluation of DML in carpal tunnel syndrome (CTS) has a pooled sensitivity of 0.62 and a specificity of 0.94 [29].
- In severe CTS, it is not unusual that the conduction velocity in the antecubital fossa-wrist segment is slowed. It should be reminded that conduction velocity reflects conduction of the largest-diameter fastest-conducting fibers that are more susceptible to compression; if these fibers are blocked at the wrist or have undergone Wallerian degeneration, only the velocity of the spared slower-conducting myelinated fibers can be measured, producing a slow conduction velocity in the forearm segment. Therefore, in a patient with severe CTS, a slowed forearm median motor conduction velocity does not imply an additional proximal trouble.
- If the amplitude of the CMAP is higher at the proximal than distal stimulation site and if the distal motor latency is prolonged because of CTS, but the conduction velocity across the forearm is higher than normal, consider a Martin-Gruber anastomosis.

4.17 Median Nerve Sensory Study to Digit 2 or 3

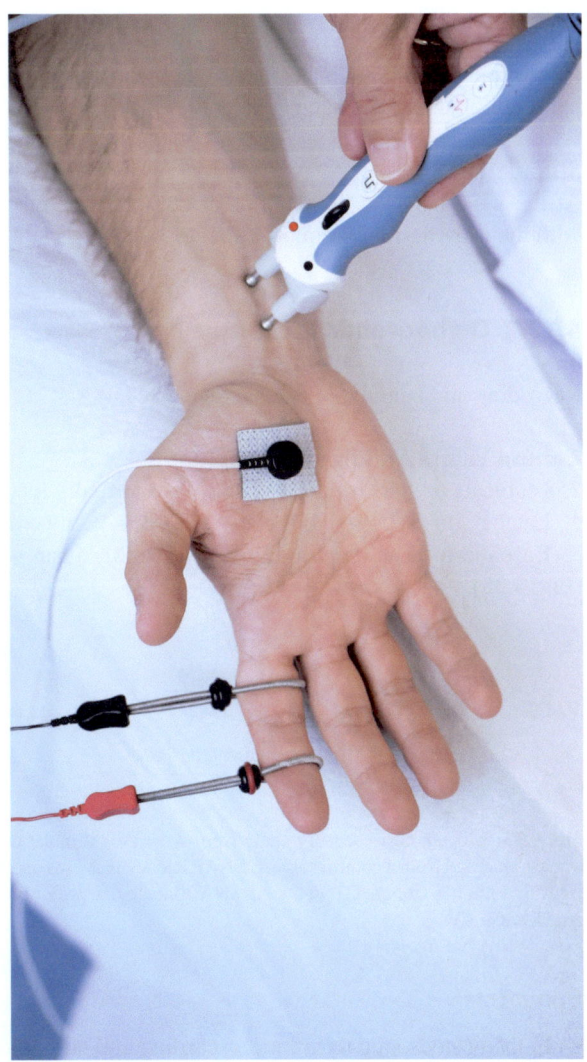

Fig. 4.34 Median nerve sensory study to digit 2

Nerve Fibers Tested and Route
Digital nerves, median nerve, lateral cord, upper and middle trunks, C6 (digit 2), C7 (digit 3).

4.17.1 Antidromic Technique

Recording Site
By using ring electrodes on digit 2 (or digit 3) with A placed over the metacarpophalangeal joint and R 3–4 cm distally over the interphalangeal joint. G is placed on the palm or dorsum of the hand (Fig. 4.34).

Stimulation Site
At the wrist between the tendons of the flexor carpi radialis and palmaris longus with the cathode 13–14 cm proximal to A (Fig. 4.34).

4.17.2 Orthodromic Technique

Recording and stimulation sites are reversed.

Control Values Antidromic Technique
258 subjects (results for digits 2 and 3 are virtually identical) [30].
   Latency (ms), 2.7 ± 0.3; mean + 2SD, 3.3; 97th percentile 3.2.
   Baseline-to-peak amplitude (μV), 41 ± 20; mean −2SD, 10; 3rd percentile 14 (Fig. 4.35).

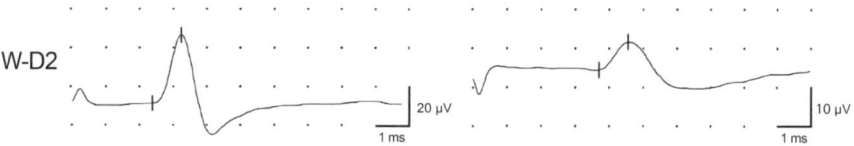

Fig. 4.35 Median nerve sensory study from wrist (W) stimulation to digit 2 (D2). On the left is a tracing recorded from a control subject: SNAP latency is 2.7 ms and SNAP amplitude is 34 μV. On the right is a tracing from a CTS patient: SNAP latency is prolonged to 3.7 ms, SNAP amplitude is reduced to 7 μV

Comments
- In the orthodromic technique, recording and stimulation sites are reversed.
- In the orthodromic technique, the SNAP has a lower amplitude.
- In the antidromic technique, if the stimulus artifact obscures the onset of the SNAP, rotate the anode of the stimulator by 45°. An excessive stimulus artifact affects the measurement of the SNAP latency and amplitude (Fig. 4.36).

4.17 Median Nerve Sensory Study to Digit 2 or 3

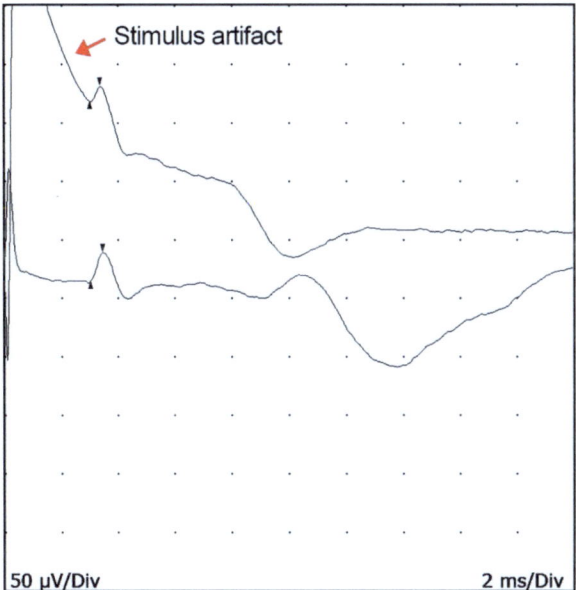

Fig. 4.36 Median nerve sensory study to digit 2, control subject. In the upper tracing with normal orientation of the stimulator a stimulus artifact is present. Rotating the anode of the stimulator by 45° greatly reduces the stimulus artifact (lower tracing). The negative stimulus artifact in the upper tracing may factitiously increase the onset latency and decrease the amplitude of SNAP

- In the antidromic technique also, the median nerve motor fibers are stimulated, and a volume-conducted motor potential (motor artifact, originating from the median-innervated hand muscles) with a slightly longer latency may be recorded (Fig. 4.37). If the SNAP is delayed, as in CTS, the motor artifact may obscure SNAP measurements. If this occurs, reposition the active and recording electrodes more distally on the digit to reduce the motor artifact (Fig. 4.37). Caveat: in case of an absent SNAP, the motor artifact might be misinterpreted as a delayed SNAP. If in doubt, perform an orthodromic study that is not disturbed by the motor artifact.

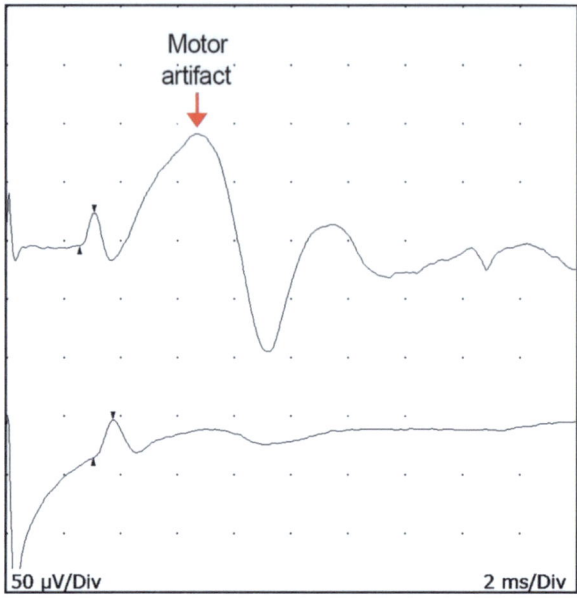

Fig. 4.37 Median nerve sensory study to digit 2, control subject. In the upper tracing the SNAP is followed by a volume-conducted motor potential (motor artifact). In the lower tracing, after repositioning the active and recording electrodes more distally on the digit, the motor artifact is reduced

- Stimulation can be performed also proximally at the antecubital fossa, at the same site of the median motor study. The proximal SNAP may be considerably smaller in amplitude because of physiological temporal dispersion and phase cancellation.
- In a review study, the evaluation of median sensory conduction in CTS has a pooled sensitivity of 0.62 and a specificity of 0.94 [31].
- Median sensory nerve study is normal in anterior interosseous neuropathy.

4.18 Median to Ulnar Comparative Studies

In patients with CTS, the median DML, the sensory latency, and the sensory conduction velocity are usually moderately to markedly prolonged. However, in up to 25% of patients with typical CTS symptoms, these routine studies may be normal. In this case, more sensitive studies involving a comparison of the median nerve to another nerve (ulnar or radial) in the same hand are useful. The most used comparison tests are as follows: (1) median versus ulnar digit 4 sensory study, (2) median versus ulnar palm-wrist mixed nerve study, and (3) median second lumbrical versus ulnar interossei DML study. These techniques utilize an ideal internal control in which several variables that affect nerve conduction velocity, such as age, distance, and temperature, are held constant. The only factor that varies is that the median nerve passes through the carpal tunnel, whereas the ulnar nerve does not. Thus, any conduction slowing of the median nerve, compared with the ulnar, can be ascribed to the passage across the carpal tunnel.

Moreover, all the above tests give information on ulnar nerve conduction which should always be investigated in diagnosing CTS in order to exclude a peripheral neuropathy.

4.18.1 Median Versus Ulnar Digit 4 Sensory Study

In the great majority (about 97%) of subjects, the sensory innervation to the ring finger is split, with the lateral half supplied by the median nerve and the medial half by the ulnar nerve.

4.18.1.1 Antidromic Technique

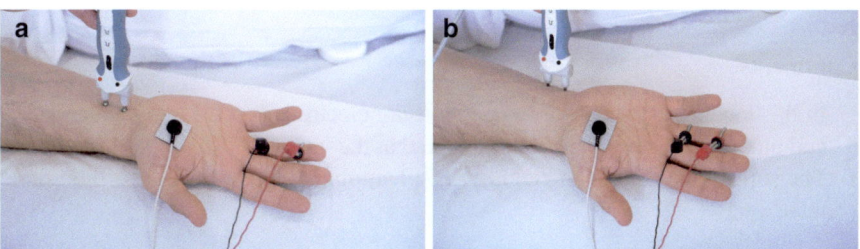

Fig. 4.38 Antidromic median and ulnar digit 4 sensory study

Recording Site
By using ring electrodes with A placed over the proximal interphalangeal joint and R placed 3–4 cm distally over the distal interphalangeal joint of digit 4. G is placed on the palm or dorsum of the hand (Fig. 4.38).

Stimulation Sites
Median nerve at the wrist: slightly lateral to the mid-wrist between the tendons of the flexor carpi radialis and palmaris longus (Fig. 4.38a). Ulnar nerve at the wrist: adjacent to the flexor carpi ulnaris tendon (Fig. 4.38b). The same distance (14 cm) must be used for both studies.

Control Values
60 subjects, 100 hands [32].
  ULN of median to ulnar D4 sensory latency difference: 0.4 ms (Fig. 4.39).

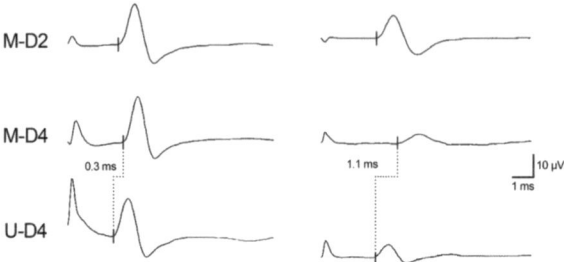

Fig. 4.39 Antidromic median and ulnar digit 4 sensory study. On the left are tracings recorded from a control subject, on the right are tracings from a CTS patient. SNAP recorded over digit 2 (D2) and digit 4 (D4) after the median (M) and ulnar (U) nerve stimulation. M-D2 latency is in the normal range both in the control (2.6 ms) and in the CTS patient (2.8 ms). The latency difference between M-D4 and U-D4 is 0.3 ms in the control and increased to 1.1 ms in the CTS patient

Comments

- In this technique, using the same recording electrodes on D4, the median nerve and ulnar nerve are separately stimulated at the wrist.
- The antidromic technique yields a SNAP amplitude larger than the orthodromic technique.

4.18.1.2 Orthodromic Technique

Fig. 4.40 Orthodromic median and ulnar digit 4 sensory study

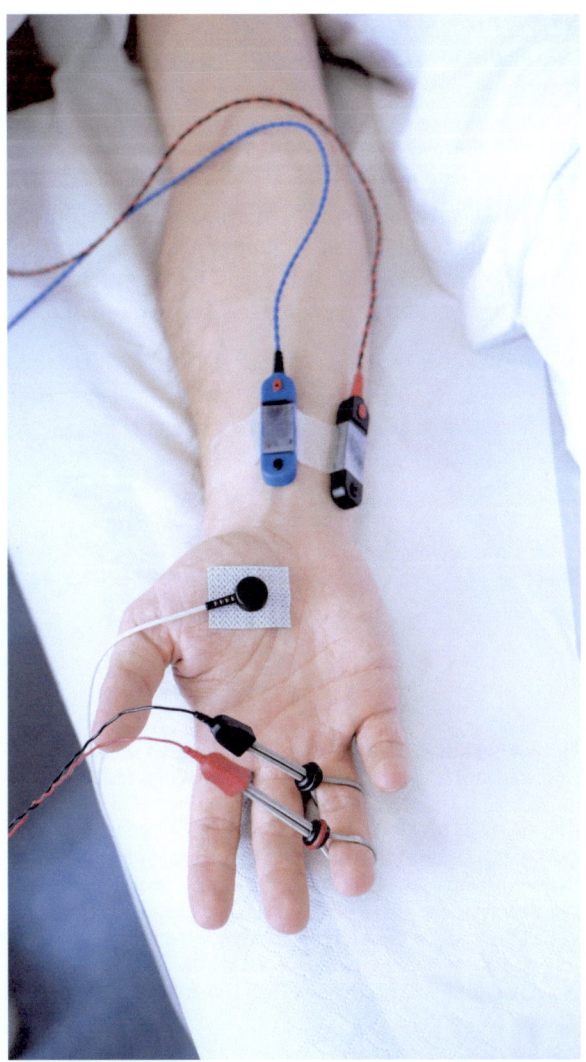

Recording Sites
Dual channel recording by two 3 cm bar electrodes (A distal, R proximal) placed longitudinally and taped at the wrist over the median nerve (blue bar, between the tendons of the flexor carpi radialis and palmaris longus) and ulnar nerve (black bar, adjacent to the flexor carpi ulnaris tendon). G is placed on the palm or dorsum of the hand (Fig. 4.40).

Stimulation Site

By using ring electrodes with the cathode placed over the proximal interphalangeal joint and the anode 3–4 cm distally over the distal interphalangeal joint of digit 4 (Fig. 4.40).

The distance between the stimulating cathode and A must be identical at the two recording sites (13–14 cm).

Control Values

47 controls, 72 hands [33].

ULN of median to ulnar D4 sensory latency difference (mean ± 2SD and coinciding with the maximum of range values): 0.4 ms (Fig. 4.41).

ULN of median to ulnar D4 sensory CV difference (mean ± 2.5SD): 9.0 m/s [34].

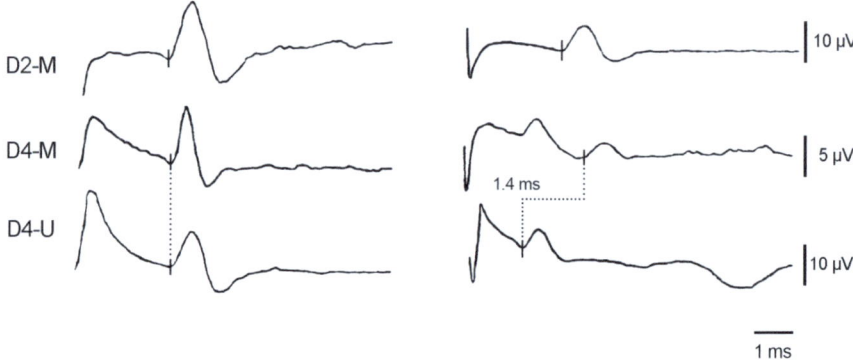

Fig. 4.41 Orthodromic median and ulnar digit 4 sensory study. On the left are tracings recorded from a control subject, and on the right are tracings from a CTS patient. The SNAP is recorded over the median (M) and ulnar (U) nerve after stimulation of digit 2 (D2) and digit 4 (D4). In the control, the D2-M SNAP latency is normal (2.6 ms), and there is no difference between the D4-M and D4-U latencies. In a CTS patient, the D2-M latency is still in the normal range (2.9 ms). Stimulation of D4 produces a double-peak potential, "the camel sign," recorded over the median nerve (D4-M). The shorter latency peak is a volume-conducted ulnar SNAP as it has a lower amplitude, but the same latency of the SNAP is recorded over the ulnar nerve after D4 stimulation (D4-U), and it is followed by the delayed median SNAP. The latency difference between D4-M and D4-U is increased to 1.4 ms. Reproduced from Uncini et al. 1989 with permission by Wiley [35]

Comments

- Using the same stimulating electrodes at D4 and two recording channels, median and ulnar SNAPs can be contemporaneously recorded at the wrist reducing the number of stimuli delivered to the subject and the examination time.
- The orthodromic technique has the additional advantage of producing the double-peak potential (also known as the camel sign) that is observed when the D4 median to ulnar latency difference exceeds 0.7 ms. [35] (Fig. 4.42).

4.18 Median to Ulnar Comparative Studies

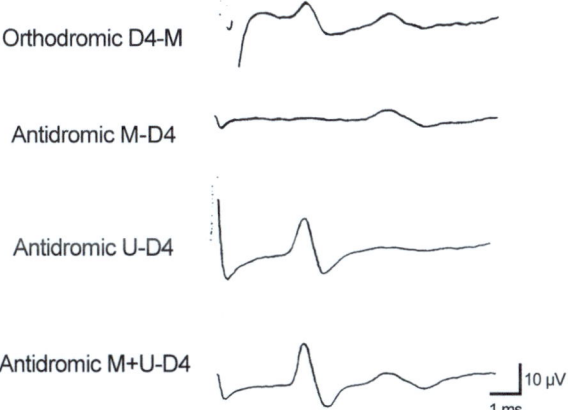

Fig. 4.42 Demonstration of the sources of the double-peak potential in a CTS patient. Upper tracing: orthodromic sensory conduction from digit 4 to median (D4-M). Middle tracings: selective median (M) or ulnar (U) antidromic conductions from the wrist to D4. Lower tracing: combined median and ulnar (M+U) antidromic conduction from the wrist to D4. Note that antidromic stimulation of the median nerve produces only the longer latency peak, while antidromic stimulation of the ulnar nerve produces only the shorter latency one. Combined stimulation of both nerves at the wrist reproduces antidromically the double-peak response as in orthodromic recording. Reproduced from Uncini et al. 1990 with permission by Elsevier [34]

- Sometimes the volume-conducted ulnar potential recorded at the median site may obscure the true median potential onset resulting in an artifactually prolonged latency. Nevertheless, the double-peak potential is the direct visual demonstration of the conduction delay of median sensory fibers across the carpal tunnel with immediate diagnostic implications.
- One objection to the orthodromic technique is that when only one potential is seen over the median nerve, it would not be possible to differentiate between a volume-conducted ulnar potential and no recordable median potential, a median potential without a volume-conducted ulnar potential, or superimposition of both median and ulnar potentials. This objection is easily overruled by the fact that in 4th finger technique, the D4 ulnar SNAP is always recorded and helps to differentiate the above conditions.
- Moreover, it should be reminded that this comparative test should always follow median D2 sensory conduction to reduce possible pitfalls in a CTS diagnosis.
- In a review study, the comparison between median and ulnar sensory conduction between wrist and ring finger in CTS has a pooled sensitivity of 0.85 and a specificity of 0.97 [36].

4.18.2 Median Versus Ulnar Mixed Nerve Study from Palmar Stimulation

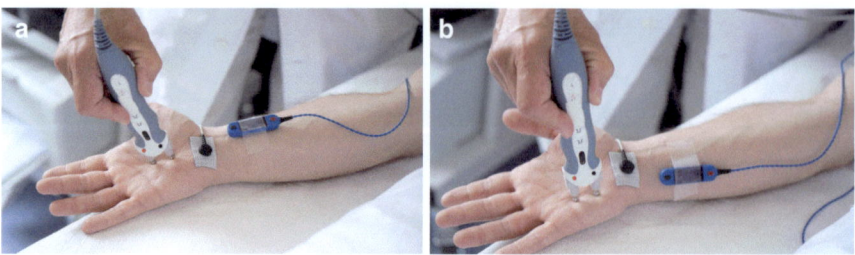

Fig. 4.43 Median (**a**) and ulnar (**b**) mixed nerve study from palmar stimulation

4.18.2.1 Median Nerve
Recording Site
Median nerve at the wrist by a 3 cm bar electrode. A is placed at the proximal wrist crease between the tendons of the flexor carpi radialis and palmaris longus, and R is proximal. G is between A and the stimulating cathode (Fig. 4.43a).

Stimulation Site
Median nerve in the palm 8 cm from A on a line drawn from the wrist to the web space between the index and middle fingers. The cathode is proximal (Fig. 4.43a).

4.18.2.2 Ulnar Nerve
Recording Site
The ulnar nerve at the wrist by a 3 cm bar electrode. A is placed at the proximal wrist crease adjacent to the flexor carpi ulnaris tendon, and R is proximal. G is between A and the stimulating cathode (Fig. 4.43b).

Stimulation Site
The ulnar nerve in the palm 8 cm from A on a line drawn from the wrist to the web space between the ring and little fingers. The cathode is proximal (Fig. 4.43b).

4.18 Median to Ulnar Comparative Studies

Control Values

47 controls, 72 hands [33].

ULN of median to ulnar mixed nerve latency difference (mean ± 2SD and coinciding with the maximum of range values): 0.4 ms (Fig. 4.44).

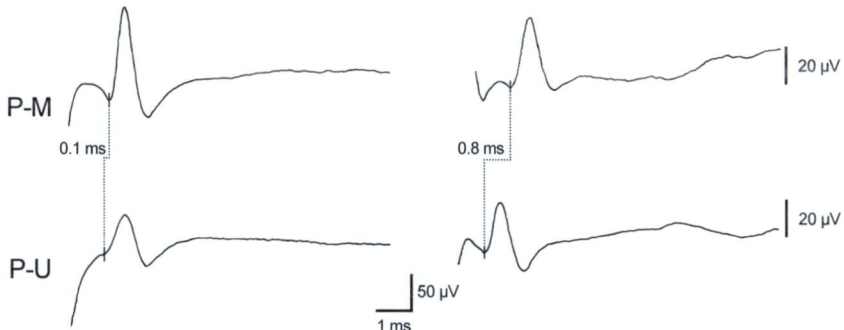

Fig. 4.44 Mixed nerve action potentials recorded at the wrist after palmar stimulation from the median (P-M) and ulnar nerve (P-U). On the left are recordings from a normal subject, and on the right are recordings from a CTS patient with normal median DML and normal W-D2 SNAP latency. In the control, the difference in the latency onset of mixed potential between P-M and P-U is 0.1 ms, whereas in the CTS patient, it is increased to 0.8 ms. Reproduced from Uncini et al. 1993 with permission by Wiley [33]

Comments

- This technique measures mixed nerve potentials at the wrist resulting from the excitation of both sensory fibers activated orthodromically and motor fibers innervating the lumbrical and interossei muscles activated antidromically.
- The amplitude of mixed potentials is considerably larger than the amplitude of SNAPs.
- The advantage of this technique is that median nerve conduction is calculated over a short distance (8 cm) with, in the case of CTS, little dilution of slowing due to the inclusion of normal nerve segments.
- In a review study, the comparison of median and ulnar mixed nerve conduction between the palm and wrist in CTS has a pooled sensitivity of 0.74 and a specificity of 0.97 [36].

4.18.3 Median 2nd Lumbrical Versus Ulnar Interossei Distal Motor Latencies Study

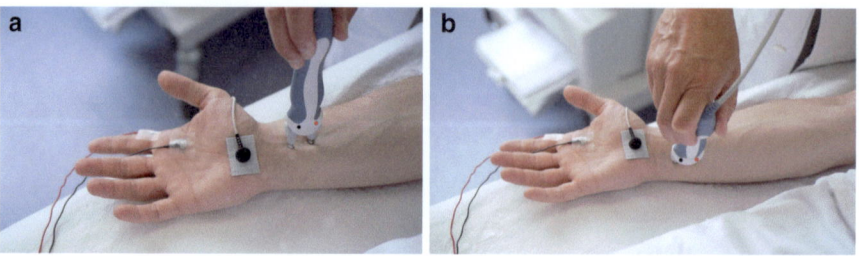

Fig. 4.45 Median 2nd lumbrical vs ulnar interossei distal motor latencies study. (A) Stimulating the median nerve at the wrist and recording from the 2nd lumbrical muscle. (B) Stimulating the ulnar nerve at the wrist and recording from interossei. Note that the same active electrode (black wire) placed lateral to the midpoint of the third metacarpal bone allows the recording of CMAPs from the 2nd lumbrical and the deeper interossei

Recording Site

The same for the 2nd lumbrical (median innervated) and interossei (ulnar innervated). A is placed slightly lateral to the midpoint of the third metacarpal bone; R is placed distally over the proximal inter phalangeal joint of digit 2. G is placed between A and the cathode of the stimulator or on the dorsum of the hand (Fig. 4.45).

Stimulation Sites

Median nerve at the wrist between the tendons of the flexor carpi radialis and palmaris longus (Fig. 4.45a); ulnar nerve at the wrist adjacent to the flexor carpi ulnaris tendon (Fig. 4.45b). The same distance (8–10 cm from A) must be used for median and ulnar stimulation.

Control Values

47 controls, 72 hands [33].

ULN of lumbrical to interossei distal motor latency difference (mean ± 2SD and coinciding with the maximum of range values): 0.5 ms (Fig. 4.46).

51 hands [37].

ULN of lumbrical to interossei distal motor latency difference (mean ± 2SD and coinciding with the maximum of range values): 0.4.

4.18 Median to Ulnar Comparative Studies

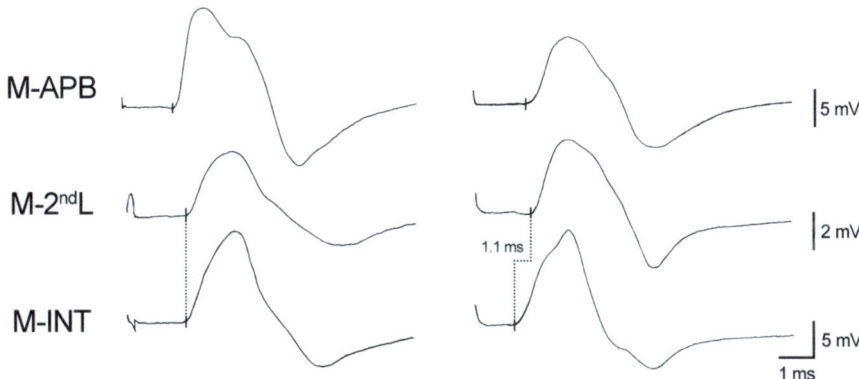

Fig. 4.46 Median 2nd lumbrical versus ulnar interossei distal motor latencies study. Left, recordings from a control subject; right, recordings from a CTS patient. Upper tracings are CMAPs recorded from the abductor pollicis brevis after stimulation of the median at the wrist (M-APB). Middle tracings are CMAPs recorded from the 2nd lumbrical after stimulation of the median at the wrist (M-2ndL). Lower tracings are CMAPs recorded from the second interossei after ulnar stimulation at the wrist (U-INT). In the control subject, M-APB latency is normal (3.1 ms), and there is no difference between M-2ndL and U-INT latencies. In CTS, the M-APB distal motor latency is equal to the control (3.1 ms), but the latency difference between M-2ndL and U-INT is increased to 1.1 ms. [32] Reproduced from Uncini et al. 1993 with permission by Wiley [32]

Comments

- The median-innervated 2nd lumbrical muscle lies just above the ulnar-innervated interossei; therefore, using the same recording electrodes, the CMAPs of the second lumbrical and the interossei can be recorded after median nerve and ulnar nerve stimulation, respectively.
- Excessive stimulation must be avoided to prevent co-stimulation of the median and ulnar nerves.
- If the 2nd lumbrical CMAP does not have a fast-rising initial negative deflection from the baseline, the active recording electrode should be repositioned.
- The interossei CMAP amplitude usually is higher than the 2nd lumbrical CMAP amplitude.
- The second lumbrical is relatively spared in CTS [38].
- In a review study, the comparison of the median 2nd lumbrical DML (second lumbrical) to ulnar interossei DML has a pooled sensitivity of 0.56 (which is the lowest of comparative tests) and a specificity of 0.98 [36].
- However, this technique is useful to demonstrate median neuropathy at the wrist in severe CTS with absent median D2 SNAP and CMAP from APB and in patients with coexistent polyneuropathy, in whom median sensory and mixed nerve potentials may be absent [39, 40].

4.19 Median Motor and Sensory Segmental Studies

In compressive neuropathies, both conduction block and axonal degeneration cause weakness and impaired voluntary recruitment of motor units. However, the prognosis for nerve recovery is quite different, as conduction block is potentially completely reversible in a relatively short time, whereas the recovery from axonal degeneration, depending on axonal regeneration, is lengthy and may be incomplete. In CTS, standard techniques do not distinguish how much of the reduced CMAP or SNAP amplitudes is due to focal demyelination and conduction block or to axonal degeneration. Conduction block and conduction slowing can be demonstrated by segmental studies comparing the amplitude of the motor and sensory responses from wrist and palm stimulation. Moreover, segmental studies focus on the conduction in the wrist-to-palm segment, excluding the normal distal segment with little dilution of slowing.

4.19.1 Motor Segmental Study

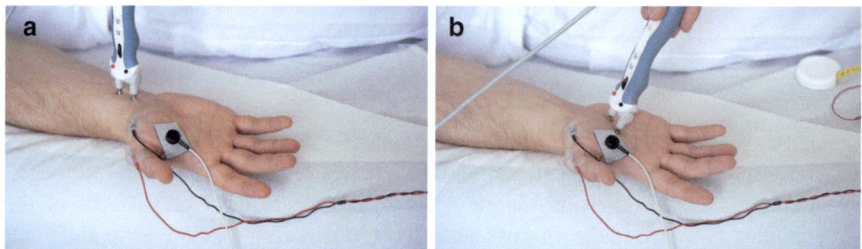

Fig. 4.47 Median motor segmental study. (**a**) Stimulation over the median nerve at the wrist and recording from the abductor pollicis brevis muscle. (**b**) Stimulation over the median nerve at the palm

Recording Site

A is placed over the muscle belly of the abductor pollicis brevis muscle halfway between the metacarpophalangeal joint of the thumb and the midpoint of the distal wrist crease, and R is placed over the first metacarpophalangeal joint. G is placed between A and the cathode of the stimulator on the dorsum of the hand (Fig. 4.47).

Stimulation Sites

At the wrist between the tendons of the flexor carpi radialis and palmaris longus 2 cm proximal to the distal wrist crease (Fig. 4.47a); at the palm 3 cm distal to the wrist crease along a line connecting the wrist to the web space between the index and middle fingers (Fig. 4.47b).

Amplitudes of the negative phase of CMAPs are measured, and the CMAP amplitude ratio between palm and wrist stimulations is calculated. The segmental

4.19 Median Motor and Sensory Segmental Studies

motor conduction velocity from the wrist to palm can be calculated by dividing the wrist-to-palm distance by the latency difference from wrist and palm stimulation.

Control Values
69 subjects, 88 hands [41].
Wrist-APB, negative peak CMAP amplitude (mV): 10.2 ± 2.9.
Palm-APB, negative peak CMAP amplitude (mV): 10.5 ± 2.9.
Wrist/palm CMAP amplitude ratio: 0.9 ± 0.1; LLN (mean −2 SD) 0.7.
Wrist/palm motor CV (m/s): 46.7 ± 5.8; LLN (mean −2 SD) 35 (Fig. 4.48).

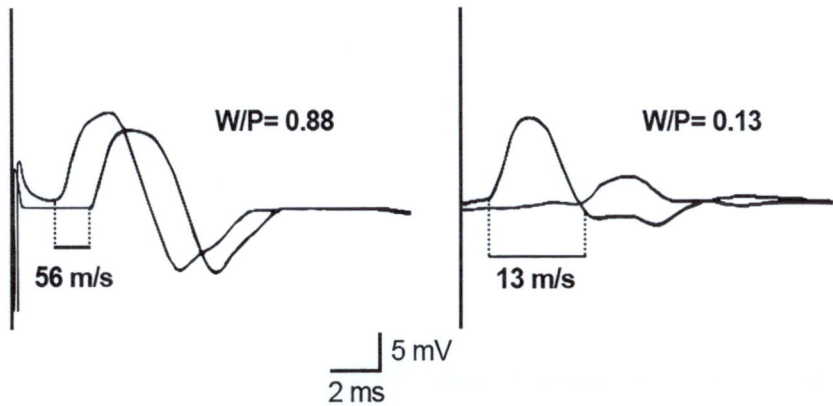

Fig. 4.48 Median motor segmental study. CMAPs after wrist and palm stimulation. Left, recordings from the control subject with normal motor CV and normal CMAP amplitude ratio in the wrist-to-palm (W/P) segment. Right, recordings from a CTS patient with slow CV and reduced CMAP amplitude ratio in the W/P segment. Reproduced from Di Guglielmo et al. 1997 with permission by Elsevier [41]

Comments

- The APB is innervated via the recurrent thenar motor branch of the median nerve, which runs into the palm and then curves back to innervate the thenar muscles.
- This technique is not easy and has some technical problems and pitfalls [41].
- Particular care should be taken to avoid the co-stimulation at the palm of the deep branch of the ulnar nerve which can also generate a motor response recordable over the APB and may result in a greater CMAP amplitude from palm than from wrist stimulation, simulating a partial conduction block. An initial positive deflection of CMAP from palm stimulation suggests co-stimulation of the deep ulnar branch. This can be avoided by (1) using proximal palmar stimulation (3 cm distal to the wrist crease), (2) increasing the stimulus at the palmar site in discrete steps as the intensity necessary to maximally stimulate the recurrent median nerve branch is always lower than the threshold for stimulation of the

deep ulnar nerve branch, (3) inspecting the thenar twitch and avoiding thumb adduction which indicates adductor pollicis contraction due to deep ulnar nerve branch co-stimulation, and (4) comparing CMAP amplitude, waveform configuration, and initial deflection from palm and wrist stimulation, as a change between these two sites suggests that a different population of fibers has been stimulated.
- Overstimulation at the palm may activate the recurrent thenar branch distally near the motor point making fictitiously greater the latency difference between the palm and wrist and resulting in an erroneously slow conduction in the wrist-to-palm segment.
- With more distal stimulation sites at the palm (more than 3 cm from the distal wrist crease), the recurrent median thenar branch may be activated under the anode, and the distance measured at the cathode would overestimate the effective nerve length making the wrist-to-palm conduction velocity erroneously fast.
- For the above reasons, the calculation of conduction velocity in the wrist/palm segment is not always reliable.
- A wrist-to-palm CMAP amplitude ratio <0.7 implies some conduction block across the wrist with a better prognosis than axonal loss.
- In a review study, the segmental motor technique has in CTS a pooled sensitivity of 0.69 and a specificity of 0.98 [42].

4.19.2 Sensory Segmental Study

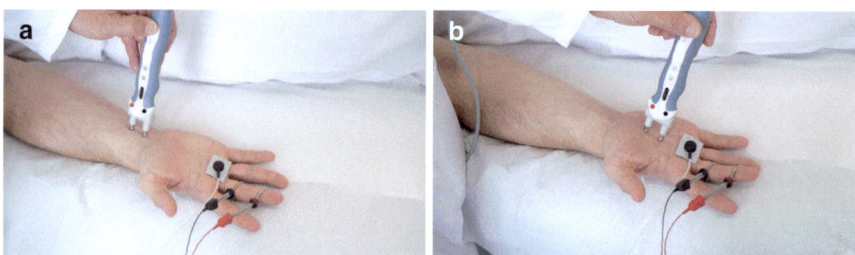

Fig. 4.49 Median nerve sensory segmental study. (**a**) Stimulation site over the median nerve at the wrist, recording by using ring electrodes from D3. (**b**) Stimulation site over the median nerve in the palm

Recording Site
D3 or D2 by using ring electrodes with A over the proximal interphalangeal joint, R 3–4 cm distally over the distal interphalangeal joint (Fig. 4.49).

Stimulation Sites
Wrist: between the tendons of the flexor carpi radialis and palmaris longus with the cathode 14 cm proximal to A (Fig. 4.49a). Palm: 7 cm distal to the wrist stimulation

site along a line connecting the wrist to the web space between the index and middle fingers (Fig. 4.49b).

Baseline-to-peak amplitude of the SNAP from wrist and palm stimulation is measured, and the SNAP amplitude ratio between palm and wrist stimulation is calculated. The segmental sensory conduction velocity from the wrist to palm is calculated by dividing the wrist-to-palm distance by the latency difference from wrist and palm stimulation.

Control Values
69 subjects, 88 hands [41].
  Wrist-D2 SNAP amplitude (baseline to peak): 42 ± 19 µV.
  Palm-D2 SNAP amplitude (baseline to peak): 45 ± 21 µV.
  Wrist/palm SNAP amplitude ratio: 0.8 ± 0.2, LLN (mean −2 SD) 0.5.
  Wrist/palm sensory CV: 58.1 ± 6.4 m/s, LLN (mean −2 SD) 45 m/s (Fig. 4.50).

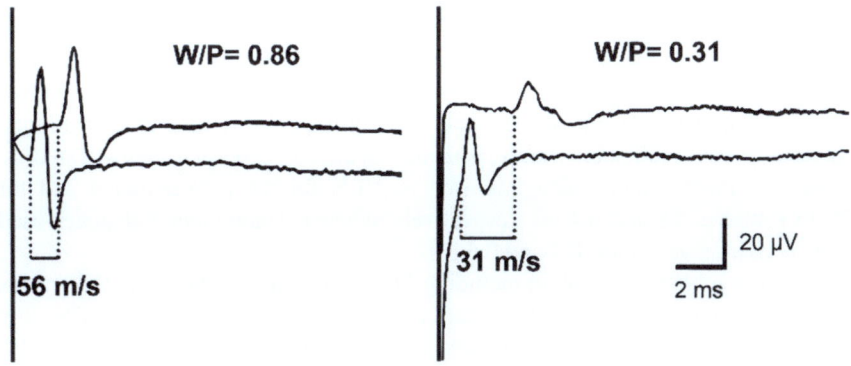

Fig. 4.50 SNAPs recorded from the index finger after wrist and palm stimulation. Left: tracings from a control subject with normal sensory CV and normal SNAP amplitude ratio in the wrist-to-palm (W/P) segment. Right: tracings from a CTS patient with slow sensory CV and reduced SNAP amplitude ratio in the W/P segment. Reproduced from Di Guglielmo et al. 1997 with permission by Elsevier [41]

Comments

- At palm stimulation, stimulus artifact may induce a baseline distortion obscuring the onset latency. In this case, the anode should be rotated until a suitable baseline is obtained.
- A wrist-to-palm SNAP amplitude ratio <0.5 indicates some conduction block across the wrist [41].
- In a review study, the segmental sensory technique has in CTS a pooled sensitivity of 0.85 and a specificity of 0.98 [42].

4.20 Yield of Comparative and Segmental Studies in CTS

Uncini and colleagues studied 193 hands from 113 consecutive patients (78% women) with clinical signs of CTS, normal median DML, and normal or borderline sensory conduction velocity from digit 2 stimulation [43]. The orthodromic difference between the median and ulnar sensory latencies from digit 4 stimulation showed the greatest sensitivity (77%), followed by the difference between the median and ulnar mixed nerve latencies from palmar stimulation (56%) and the difference between 2ndL-INT DML (10%). The median nerve intraneural funicular topography, differential compressive effects at the level of the distal carpal tunnel, and the different types of fibers stimulated in each comparative test might contribute to explaining why the digit 4 median to ulnar sensory latency difference is more sensitive than the other two tests.

Preston and colleagues compared the median to ulnar latency difference studies in 34 patients with mild CTS [44]. The palmar-mixed, digit 4, and 2ndL-INT studies were abnormal in 97%, 91%, and 88% of patients, respectively. Two or more tests were abnormal in 97% of hands. The high sensitivity found for the 2L-INT test (88%) was in contrast with the low sensitivity (10%) found by Uncini and colleagues [43]. This disparity can be explained by the different patient cohorts: Uncini's cohort was larger, and patients had normal or borderline median routine studies, whereas in Preston's study, patients had already abnormal standard studies. Moreover, Preston and colleagues used as ULN for the palmar-mixed and for 2L-INT studies 0.3 and 0.4 ms, respectively, whereas Uncini and colleagues used more conservative values: 0.4 and 0.5 ms.

In a large multicentric cohort including 1123 symptomatic idiopathic CTS hands (740 patients), the sensitivity of standard tests (median DML and sensory CV) was 83.5%. Comparative/segmental tests disclosed abnormal findings in a further 11.4% of hands increasing the overall electrodiagnostic sensitivity to 94.9% [45].

Overall, patients with early or mild CTS may have normal standard median motor and sensory studies, and rare patients will have even normal comparative/segmental tests. It is evident that in CTS no electrodiagnostic test is 100% sensitive and specific, and electrodiagnostic results can be interpreted properly only with knowledge of the clinical history and examination.

4.21 An Electrophysiological Classification of CTS

CTS Classification	All tests normal	Abnormal segmental/ comparative tests	Abnormal SCV (digit-wrist)	Abnormal DML	Absent SNAP (digit-wrist)	Absent CMAP
EXTREME					■	■
SEVERE				■	■	
MODERATE			■	■		
MILD			■			
MINIMAL		■				
NEGATIVE	■					

Fig. 4.51 Neurophysiological CTS classification: electrophysiological patterns of different categories. For the Italian version, it is suggested to use the term "mediol for "moderate" (because "moderate" does not exactly correspond to the Italian term "moderato"). For the other CTS categories, the terms suggested are as follows: "estremo" for extreme, "grave" for severe, "lieve" for mild, and "minimo" for minimal [47]

Although it has long been recognized that in individual patients with CTS, there may be little correlation between the severity of clinical symptoms or signs and the abnormalities seen on nerve conduction studies, to standardize the electrodiagnostic approach to CTS, it is important to adopt a common classification system of the different electrophysiological patterns. A neurophysiological classification including six categories has been proposed (Fig. 4.51): (1) extreme CTS, absence of median CMAP and SNAP; (2) severe CTS, absence of median SNAP and abnormal median DML; (3) moderate CTS, slowing of median sensory conduction velocity and abnormal median DML; (4) mild CTS, slowing of median sensory conduction velocity and normal median DML; (5) minimal CTS, standard negative tests with abnormal comparative or segmental tests; (6) negative CTS, patients with typical CTS symptoms and normal findings in all tests (including comparative or segmental tests) [46].

This classification has some advantages: (a) it makes possible to easily classify CTS electrophysiologically on the basis of cut-offs (normal/abnormal conduction findings and the presence/absence of evoked responses); (b) different laboratories can use their own control neurophysiological values; (c) it can be applied regardless of the neurophysiological techniques adopted by the different laboratories (i.e., orthodromic or antidromic sensory conduction velocity and different comparative or segmental studies adopted) [47].

It should be underlined that this classification, providing a standardized neurophysiological quantitative evaluation of the median nerve impairment in CTS patients, should be used in addition to and not instead of clinical evaluation.

References

Phrenic Nerve Study

1. Newson DJ. Phrenic nerve conduction in man. J Neurol Neurosurg Psychiatry. 1967;30:420–6.

Long Thoracic Nerve Study

2. Alfonsi E, Moglia A, Sandrini G, Pisoni MR, Arrigo A. Electrophysiological study of long thoracic nerve conduction in normal subjects. Electromyogr Clin Neurophysiol. 1986;26:63–7.

Suprascapular Nerve Study

3. Kraft GH. Axillary, musculocutaneous and suprascapular nerve latency studies. Arch Phys Med Rehabil. 1972;53:382–7.
4. Buschbacher RM. Manual of nerve conduction studies. New York: Demos Medical Publishing; 2000.

Axillary Nerve Study

5. Buschbacher RM. Manual of nerve conduction studies. New York: Demos Medical Publishing; 2000.
6. Kraft GH. Axillary, musculocutaneous and suprascapular nerve latency studies. Arch Phys Med Rehabil. 1972;53:382–7.
7. Zis P, Hadjivassiliou M, Rao DG. Axillary motor nerve conduction study: description of technique and provision of normative data. J Electromyogr Kinesiol. 2018;39:95–8.

Musculocutaneous Nerve Study

8. Buschbacher RM. Manual of nerve conduction studies. New York: Demos Medical Publishing; 2000.
9. Kraft GH. Axillary, musculocutaneous and suprascapular nerve latency studies. Arch Phys Med Rehabil. 1972;53:382–7.

Lateral Cutaneous Nerve of the Forearm Study

10. Spindler HA, Felsenthal G. Sensory conduction in the musculocutaneous nerve. Arch Phys Med Rehabil. 1978;59:20–3.

Radial Nerve Motor Study

11. Preston DC, Shapiro S. Electromyography and neuromuscular disorders. Clinical electrophysiological correlations. 3rd ed. Elsevier Saunders; 2013.

Posterior Cutaneous Nerve of the Forearm Study

12. Ma DM, Liveson JA. Nerve conduction handbook. Philadelphia: F.A. Davis Co; 1983. p. 79–81.

Radial Nerve Sensory Study

13. MacKenzie K, DeLisa JA. Determining the distal sensory latency of the superficial radial nerve in normal adult subjects. Arch Phys Med Rehab. 1981;62:31–4.

Medial Cutaneous Nerve of the Forearm Study

14. Kimura J, Ayyar DR. Sensory nerve conduction study in the medial antebrachial cutaneous nerve. Tohoku J Exp Med. 1984;142:461–6.

Ulnar Nerve Motor Study

15. Busbacher RM. Ulnar nerve motor conduction to the abductor digiti minimi. Am J Phys Med Rehab. 1999;78:S9–S14.
16. Campbell WW, Pridgeon RM, Sahni KS. Short segment incremental studies in the evaluation of ulnar neuropathy at the elbow. Muscle Nerve. 1992;15:1050–4.
17. Kanakamedala RV, Simon DG, Porter RW, Zucker RS. Ulnar nerve entrapment at the elbow localized by short segment stimulation. Arch Phys Med Rehabil. 1988;69:959–63.

Deep Ulnar Motor Branch Study

18. Olney RK, Wilbourn AJ. Ulnar nerve conduction of the first dorsal interosseous muscle. Arch Physical Med Rehabil. 1985;66:16–8.
19. Buschbacher RM. Manual of nerve conduction studies. New York: Demos Medical Publishing; 2000.
20. Olney RK, Hanson M. AAEE case report #15: ulnar neuropathy at or distal to the wrist. Muscle Nerve. 1988;11:828–32.
21. Stewart JD. The variable clinical manifestation of ulnar neuropathies at the elbow. J Neurol Neurosurg Psychiatry. 1987;50:252–8.
22. Caliandro P, Foschini M, Pazzaglia C, et al. IN-RATIO: a new test to increase diagnostic sensitivity in ulnar nerve entrapment at elbow. Clin Neurophysiol. 2008;119:1600–6.

Ulnar Nerve Sensory Study to Digit 5

23. Kimura J. Electrodiagnosis in diseases of nerve and muscle. 3rd ed. New York: Oxford University Press; 2001.
24. Falco FJE, Hennessey WJ, Braddon RL, Goldberg G. Standardized nerve conduction studies in the upper limb of healthy elderly. Am J Phys Med Rehabil. 1992;71:263–71.

Ulnar Dorsal Cutaneous Nerve Study

25. Jabre JF. Ulnar nerve lesion at the wrist: new technique for recording from the sensory dorsal branch of the ulnar nerve. Neurology. 1980;30:873–6.
26. Stewart JD. The variable clinical manifestation of ulnar neuropathies at the elbow. J Neurol Neurosurg Psychiatry. 1987;50:252–8.

Anterior Interosseous Nerve Study

27. Craft S, Currier DP, Nelson MR. Motor conduction of the anterior interosseous nerve. Phys Ther. 1977;57:1143–7.

Median Nerve Motor Study

28. Buschbacher RM. Median nerve motor conduction study to the abductor pollicis brevis. Am J Phys Med Reahabil. 1999;78:S1–8.
29. Jablecki CK, Andary MT, Floeter MK, et al. American Association of Electrodiagnostic Medicine; American Academy of Neurology; American Academy of Physical Medicine and Rehabilitation. Practice parameters: electrodiagnostic studies in carpal tunnel syndrome. Report of the American Association of Electrodiagnostic Medicine, American Academy of Neurology, and the American Academy of Physical Medicine and Rehabilitation. Neurology. 2002;58:1589–92.

Median Nerve Sensory Study to Digit 2 or 3

30. Buschbacher RM. Median 14 and 7 cm antidromic sensory studies to digits 2 and 3. Am J Phys Med Rehab. 1999;78:S53–62.
31. Jablecki CK, Andary MT, Floeter MK, et al. American Association of Electrodiagnostic Medicine; American Academy of Neurology; American Academy of Physical Medicine and Rehabilitation. Practice parameters: electrodiagnostic studies in carpal tunnel syndrome. Report of the American Association of Electrodiagnostic Medicine, American Academy of Neurology, and the American Academy of Physical Medicine and Rehabilitation. Neurology. 2002;58:1589–92.

Median to Ulnar Comparative Studies

32. Charles N, Vial C, Chauplannaz G, Bady B. Clinical validation of antidromic stimulation of the ring finger in early diagnosis of mild carpal tunnel syndrome. Electroencephalogr Clin Neurophysiol. 1990;76:142–7.
33. Uncini A, Di Muzio A, Awad J, Manente G, Tafuro M, Gambi D. Sensitivity of three median to ulnar comparative tests in diagnosis of mild carpal tunnel syndrome. Muscle Nerve. 1993;16:1366–73.
34. Uncini A, Di Muzio A, Cutarella R, Awad J, Gambi D. Orthodromic median and ulnar fourth digit sensory conduction in mild carpal tunnel syndrome. Neurophysiol Clin. 1990;20:53–61.
35. Uncini A, Lange DJ, Solomon M, Soliven B, Meer J, Lovelace RE. Ring finger testing in carpal tunnel syndrome: a comparative study of diagnostic utility. Muscle Nerve. 1989;12:735–41.
36. Jablecki CK, Andary MT, Floeter MK, et al. American Association of Electrodiagnostic Medicine; American Academy of Neurology; American Academy of Physical Medicine and Rehabilitation. Practice parameters: electrodiagnostic studies in carpal tunnel syndrome. Report of the American Association of Electrodiagnostic Medicine, American Academy of Neurology, and the American Academy of Physical Medicine and Rehabilitation. Neurology. 2002;58:1589–92.
37. Preston DC, Logigian EL. Lumbrical and interossei recording in carpal tunnel syndrome. Muscle Nerve. 1992;15:1253–7.
38. Logigian EL, Busis NA, Berger AR, Bruyninckx F, Khalil N, Shahani BT, Young RR. Lumbrical sparing in CTS symptom: anatomic, physiologic, and diagnostic implications. Neurology. 1987;37:1499–505.

39. Boonyapisit K, Katirji B, Shapiro BE, Preston DC. Lumbrical and interossei recording in severe carpal tunnel syndrome. Muscle Nerve. 2002;25:909–13.
40. Preston DC, Ross MH, Kothari MJ, et al. The median-ulnar latency difference studies are comparable in mild carpal tunnel syndrome. Muscle Nerve. 1994;176:1469–71.

Median Motor and Sensory Segmental Studies

41. Di Gugliemo G, Torrieri F, Repaci M, Uncini A. Conduction block and segmental velocities in carpal tunnel syndrome. Electroenceph Clin Neurophysiol. 1997;105:321–7.
42. Jablecki CK, Andary MT, Floeter MK, et al. American Association of Electrodiagnostic Medicine; American Academy of Neurology; American Academy of Physical Medicine and Rehabilitation. Practice parameters: electrodiagnostic studies in carpal tunnel syndrome. Report of the American Association of Electrodiagnostic Medicine, American Academy of Neurology, and the American Academy of Physical Medicine and Rehabilitation. Neurology. 2002;58:1589–92.

Yield of Comparative and Segmental Studies in CTS

43. Uncini A, Di Muzio A, Awad J, Manente G, Tafuro M, Gambi D. Sensitivity of three median to ulnar comparative tests in diagnosis of mild carpal tunnel syndrome. Muscle Nerve. 1993;16:1366–73.
44. Preston DC, Ross MH, Kothari MJ, et al. The median-ulnar latency difference studies are comparable in mild carpal tunnel syndrome. Muscle Nerve. 1994;176:1469–71.
45. Padua L, Giannini F, Girlanda P, Insola A, Luchetti R, Lo Monaco M, Padua R, Uncini A, Tonali P. Usefulness of segmental and comparative tests in the electrodiagnosis of carpal tunnel syndrome: the Italian multicenter study. Italian CTS Study Group. Ital J Neurol Sci. 1999;20:315–20.

An Electrophysiological Classification of CTS

46. Padua L, Lo Monaco M, Gregori B, Valente EM, Padua R, Tonali P. Neurophysiological classification and sensitivity in 500 carpal tunnel syndrome hands. Acta Neurol Scand. 1997;96:211–7.
47. Padua L, Lo Monaco M, Padua R, Gregori B, Tonali P. Neurophysiological classification of carpal tunnel syndrome: assessment of 600 symptomatic hands. Ital J Neurol Sci. 1997;18:145–50.

Lower Limb Nerve Studies 5

5.1 Femoral Nerve Study

Fig. 5.1 Femoral nerve study

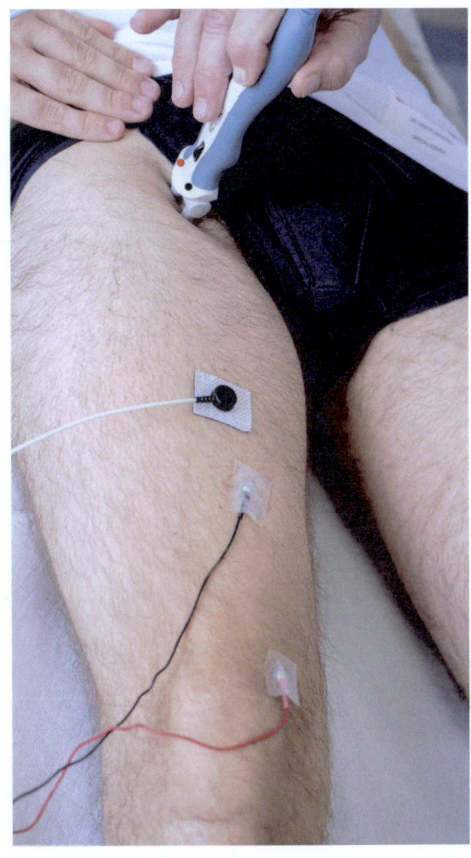

Nerve Fiber Tested and Route
L2, L3, and L4 roots, lumbosacral plexus, femoral nerve.

Recording Site
The active electrode (A) is over the vastus medialis muscle four to five fingerbreadths proximal to the superior medial angle of the patella. The reference electrode (R) is placed about 10 cm distally to A. The ground electrode (G) is placed between the stimulating cathode and the active electrode (Fig. 5.1).

Stimulation Site
The cathode is located below the inguinal ligament lateral to the point of pulsation of the femoral artery (Fig. 5.1).

Control Values
68 subjects [1].
  Age < 40 (years):
  *Latency (ms) 5.2 ± 0.5, upper limit of normal (ULN) (95th percentile) 6.0.
  Amplitude (mV) 12.1 ± 5.1, lower limit of normal (LLN) (5th percentile) 3.7.
  Age > 40 (years):
  *Latency (ms) 5.5 ± 0.5, ULN (95th percentile) 6.3.
  Amplitude (mV) 9.3 ± 5.2; LLN (5th percentile) 0.8.
  *Corrected to a standard distance of 30 cm (Fig. 5.2).

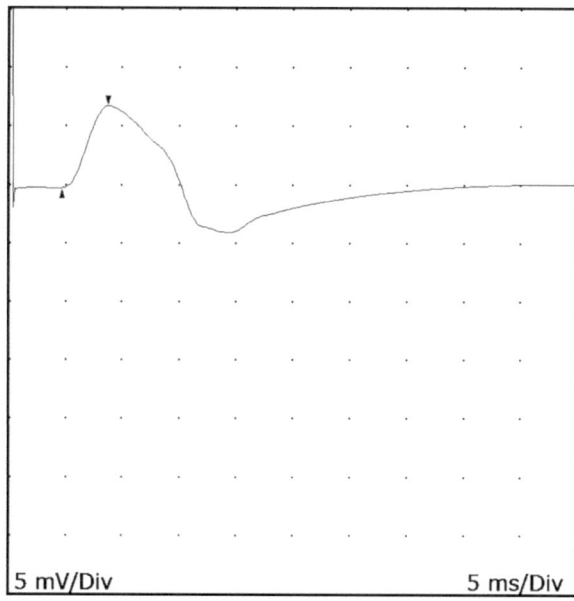

Fig. 5.2 Femoral nerve study, control subject: DML is 4.7 ms, CMAP amplitude is 6.9 mV

Comments

- With this test, the distal motor latency (DML) and the distal compound muscle action potential (CMAP) amplitude are assessed.
- Latency measurements using two stimulation points above and below the inguinal ligament may be erroneous because of the too short distance, and there is no reliable method of assessing conduction velocity (CV) in this segment.
- High current intensity and duration stimuli are needed in overweight subjects to ensure a supramaximal stimulation.
- Compare side-to-side CMAP amplitudes to assess the degree of axonal loss in femoral neuropathies, lumbar plexopathies, and severe L4 motor radiculopathies.
- The femoral nerve may be damaged during a variety of intra-abdominal, intrapelvic, and inguinal areas and hip surgical procedures.
- In hemophilia, other coagulopathies, or as a complication of anticoagulant therapy, a hematoma may form in the iliacus compartment compressing the femoral nerve.
- Femoral nerve involvement may be the most evident manifestation of diabetic radiculoplexopathy. However, in this case, EMG evidence of roots and lumbosacral plexus involvement can usually be found.

5.2 Lateral Femoral Cutaneous Nerve Study

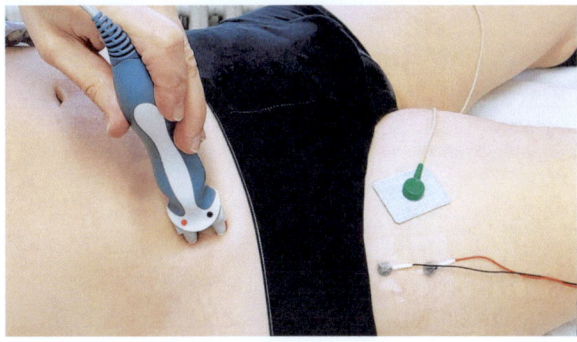

Fig. 5.3 Lateral femoral cutaneous nerve study

Nerve Fibers Tested and Route
Lateral femoral cutaneous nerve, posterior division of the lumbosacral plexus, L2 and L3 roots.

Recording Site
A is over the lateral thigh at a distance of 12 cm from the stimulating cathode. R is placed 3 cm distally from A. Ground (G) is placed between the stimulating cathode and the active electrode (Fig. 5.3).

Stimulation Site
The cathode is located one fingerbreadth medial to the anterior superior iliac spine above the inguinal ligament (Fig. 5.3).

Normal Values
58 subjects [2].
   Latency (ms): 1.7 ± 0.23.
   Amplitude (mV): 10.5 ± 4.0 (Fig. 5.4).

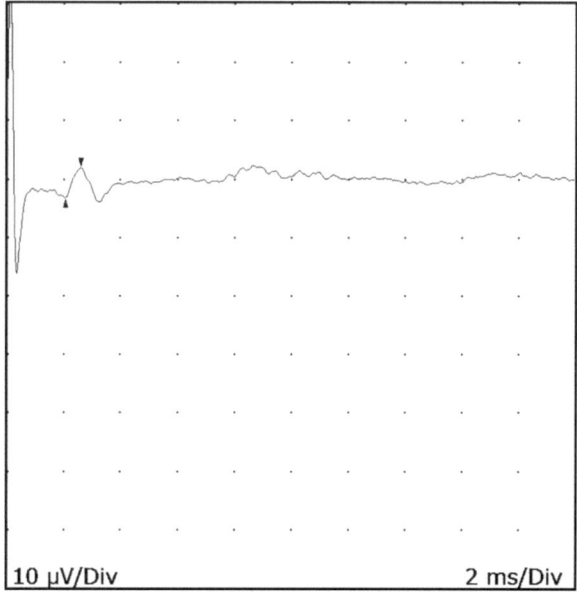

Fig. 5.4 Lateral femoral cutaneous nerve study, control subject: SNAP latency is 2.1 ms, SNAP amplitude is 5.4 μV

Comments

- The study is often technically difficult, above all in overweight subjects.
- No recordable responses are of doubtful significance.
- Side-to-side comparison is mandatory.
- It is advisable to start with the healthy side.
- May be abnormal in patients with meralgia paresthetica or lumbar plexus lesions.

5.3 Saphenous Nerve Study

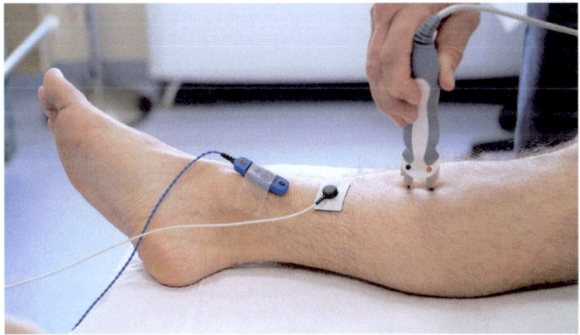

Fig. 5.5 Saphenous nerve study

Nerve Fibers Tested and Route
Saphenous nerve, femoral nerve, posterior division of lumbosacral plexus, L3 and L4 roots.

Recording Site
By a 3 cm bar electrode positioned at the medial/anterior ankle. A is placed between the medial malleolus and tibialis anterior tendon, and R is 3 cm distally. G is between the stimulating and recording sites (Fig. 5.5).

Stimulation Site
The cathode is placed 14 cm proximal to A at the medial border of the tibia. The anode is proximal (Fig. 5.5).

Control Values
80 subjects [3].
  Latency (ms), 2.9 ± 0.4; mean + 2SD, 3.7; range 2.1–3.8.
  Amplitude (µV): 5.4 ± 2.5; range 1–15 (Fig. 5.6).

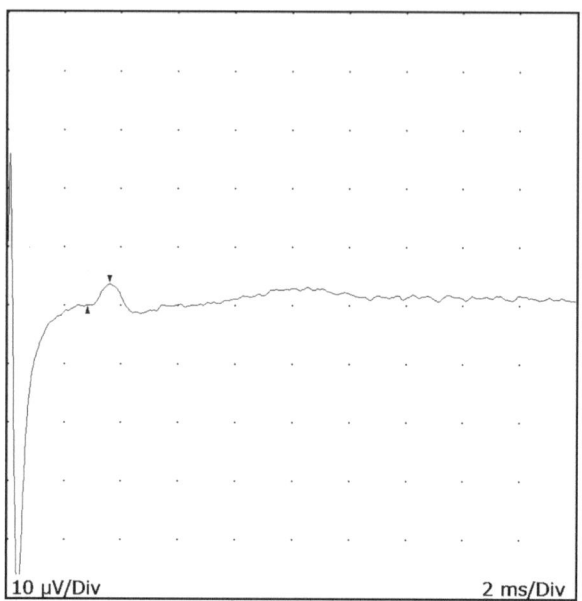

Fig. 5.6 Saphenous nerve study, control subject: SNAP latency is 2.8 ms, SNAP amplitude is 4.1μV

Comments

- To better stimulate the nerve, firmly press the stimulator against the skin.
- Low amplitude and even absent sensory nerve action potentials (SNAPs) are common especially in subjects older than 40 years.
- To maximize the response, the recording electrodes may have to be repositioned either slightly medially or laterally.
- Side-to-side comparisons of amplitude and latency are recommended.
- This study helps to differentiate lesions of the L4 root proximal to the dorsal root ganglion (normal SNAP) from lesions distal to the ganglion such as plexopathies or femoral neuropathies (abnormal SNAP).
- Isolated lesions of the saphenous nerve can occur in the thigh during femoral artery surgery such as femoral-popliteal bypass, at the knee during arthroscopy or surgical procedures, and in the lower leg during surgery for varicose veins or remotion of the saphenous vein for arterial graft.

5.4 Peroneal Nerve Study Recording from the Extensor Digitorum Brevis Muscle

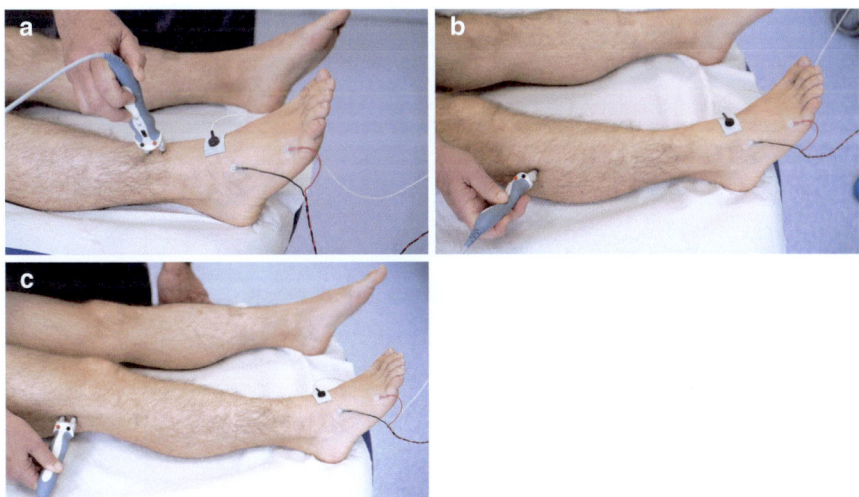

Fig. 5.7 Peroneal nerve study recording from the extensor digitorum brevis muscle. (**a**) stimulation at ankle, (**b**) stimulation below fibular head, (**c**) stimulation at popliteal fossa

Nerve Fibers Tested and Route
L5 and S1 roots, posterior division of lumbosacral plexus, sciatic nerve, common peroneal nerve, deep peroneal nerve.

Recording Site
Extensor digitorum brevis muscle, the active electrode (A) is placed over the muscle belly, the reference electrode (R) is placed over the metatarsophalangeal joint of the little toe, and the ground electrode (G) is placed on the dorsum of the foot (Fig. 5.7).

Stimulation Sites
At the ankle (A), between the extensor digitorum longus tendon and the extensor hallucis longus tendon just lateral to the tibialis anterior tendon, almost halfway between the two malleoli and 8 cm proximal to the active electrode; one to two fingerbreadths below the fibular head (BHF); at the lateral popliteal fossa (PF) near the semitendinosus muscle tendon, at a distance of 10–12 cm from below the fibular head site (Fig. 5.7).

Control Values
60 subjects, 120 nerves [4].
DML (ms): 3.77 ± 0.86; ULN (mean + 2SD) 5.5.

Distal amplitude (mV): 5.1 ± 2.3; LLN (based on the distribution of normative data) 2.5.
CV (m/s):
BHF-A: 48.3 ± 3.9; LLN (mean − 2SD) 40.
PF-BHF: 52.0 ± 6.2; LLN (mean − 2SD) 40 (Fig. 5.8).

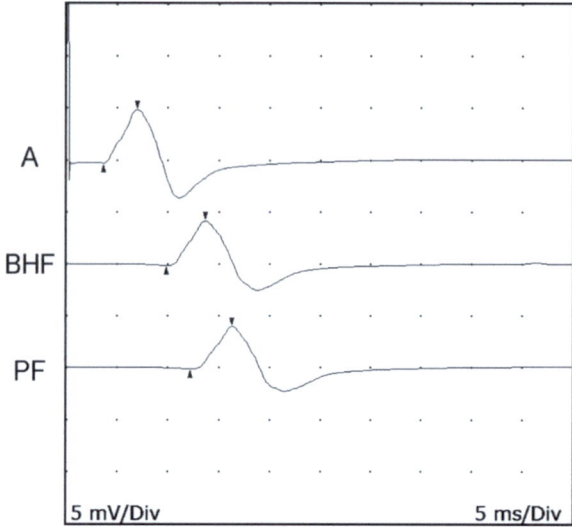

Fig. 5.8 Peroneal motor study recording from the extensor digitorum brevis muscle with stimulation at ankle (A), below fibular head (BHF) and popliteal fossa (PF), control subject: DML is 3.7 ms, distal CMAP amplitude is 5.2 mV, CV in the BHF-A segment is 51 m/s, CV in the PF-BHF segment is 50 m/s

Comments

- The peroneal nerve is deeper below the fibular head. Be careful that a supramaximal stimulation is achieved.
- Too high intensity of stimulation at the popliteal fossa can co-stimulate the tibial nerve.
- If the amplitude of CMAP is higher at the proximal than that at the distal stimulation site, consider an accessory peroneal nerve (about 28% of subjects). The presence of an accessory deep peroneal nerve is confirmed if by stimulation behind the lateral malleolus, a CMAP is recorded from the extensor digitorum brevis.

- Peroneal neuropathy at the fibular neck can be due to various conditions such as trauma, compression (cast, ganglia, etc.), and positional (squatting, leg crossing, prolonged immobilization, etc.).
- Always perform the stimulation at three points. If only the ankle and PF stimulations are done, conduction slowing across the fibular neck can be missed.
- In acute compressive neuropathies, it is necessary to wait at least 7 days after onset to establish how much of the proximal CMAP amplitude reduction is due to conduction block or conduction failure due to axonal transection (Fig. 5.9) [5].
- In acute compressive neuropathies, slowed CV across the fibular head present in the first few days after onset is due to conduction block or axonal transection of fast-conducting fibers that do not conduct from proximal stimulation but are still excitable after distal stimulation (Fig. 5.9) [5].

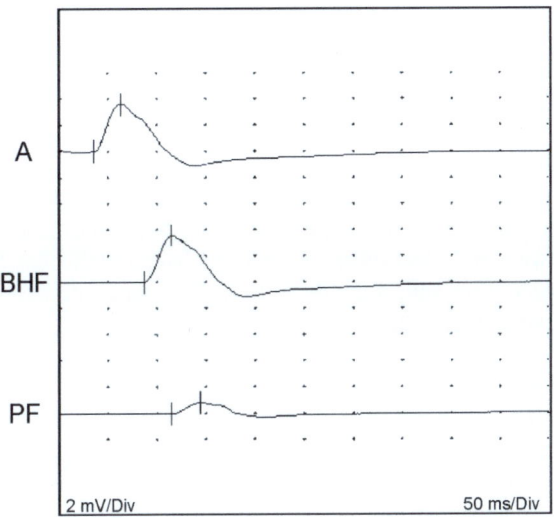

Fig. 5.9 Right peroneal nerve study from the extensor digitorum brevis muscle with stimulation at ankle (A), below fibular head (BHF) and popliteal fossa (PF) in a patient with right foot drop after prolonged leg crossing (3 days after onset). DML is 3.5 ms, distal CMAP amplitude is 3.6 mV, CV in the BHF-A segment is 51.9 m/s, CV in the PF-BHF segment is 34 m/s. The amplitude of CMAP from PF stimulation is 78% reduced compared with the CMAP from BHF stimulation indicating at this time either a conduction block or conduction failure due to axonal transection

5.5 Peroneal Nerve Study Recording from the Tibialis Anterior Muscle

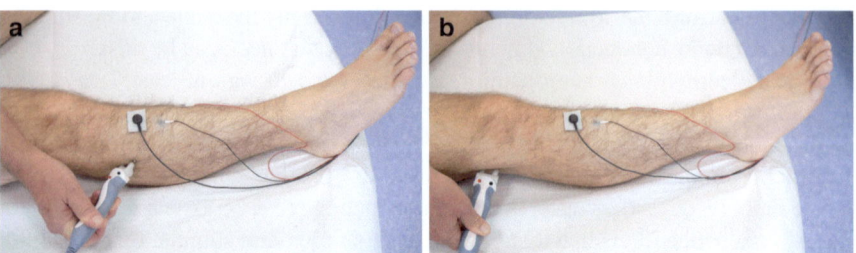

Fig. 5.10 Peroneal nerve study recording from the tibialis anterior muscle. (**a**) stimulation at the below fibular head, (**b**) stimulation at popliteal fossa

Nerve Fibers Tested and Route L4 and L5 roots, posterior division of lumbosacral plexus, sciatic nerve and common peroneal nerve.

Recording Site
The tibialis anterior muscle, in the reported technique, a 32 mm disc electrode A is placed over the muscle belly at the junction of the upper third and lower two-thirds of the line between the tibial tuberosity and the lateral malleolus. R is placed over the medial aspect of the tibia 4 cm distal to A. G is placed between the stimulating and recording sites (Fig. 5.10).

Stimulation Sites
Distal stimulation is with the cathode placed slightly below the head of the fibula; proximal stimulation is with the cathode placed just medial to the lateral border of the popliteal fossa approximately 10 cm proximal to the below fibular head stimulation site. The anode is proximal at both stimulation sites (Fig. 5.10).

Control Values
34 subjects [6].
  DML (ms): 3.0 ± 0.6, ULN (mean + 2SD) 4.2, range 2.0–4.4.
  Distal amplitude (mV): 3.9 ± 1.2.
  CV (m/s): 66.3 ± 12.9, LLN (mean-2SD) 40.5.
  Similar values are found using standard surface electrodes [7] (Fig. 5.11).

5.5 Peroneal Nerve Study Recording from the Tibialis Anterior Muscle

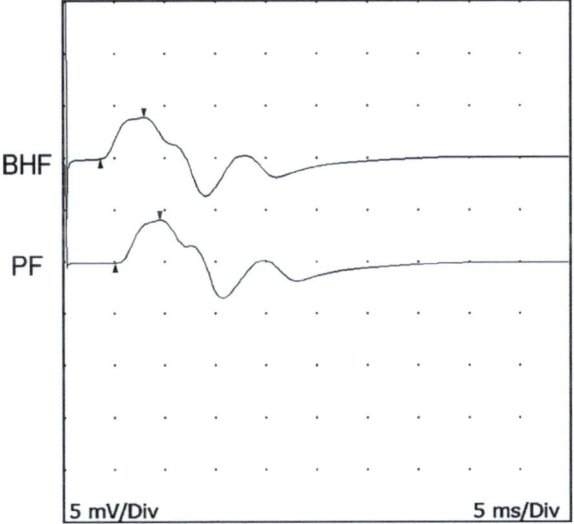

Fig. 5.11 Peroneal nerve study recording from tibialis anterior muscle, control subject. Stimulation of the peroneal nerve below fibular head (BHF), and at popliteal fossa (PF): DML is 3.6 ms, distal CMAP amplitude is 4.5 mV, CV is 65 m/s

Comments

- In peroneal nerve motor studies of the tibialis anterior, the placement of the reference electrode varies among different laboratories. The optimal placement site for the reference electrode is the tibia surface, 4 cm below the active electrode [8].
- The nerve is deeper below the fibular head. Be careful that a supramaximal stimulation is achieved.
- Too high intensity of stimulation at the popliteal fossa can co-stimulate the tibial nerve.
- This test can be useful in patients with suspected peroneal neuropathy at the fibular neck, in a foot amputee, or in cases in which more distal recording is unreliable because the extensor digitorum brevis muscle is completely or severely denervated.
- In some cases of peroneal neuropathy at the fibular neck, conduction block may be seen recording from the tibialis anterior but not from the extensor digitorum brevis.

5.6 Superficial Peroneal Sensory Study

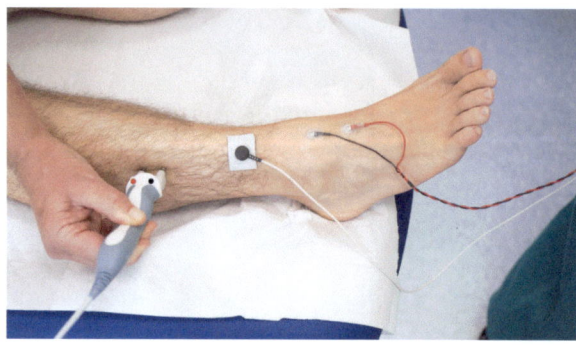

Fig. 5.12 Superficial peroneal sensory study

Nerve Fibers Tested and Route
Peroneal superficial, common peroneal nerve, posterior division lumbosacral plexus, and L4, L5, and S1 roots.

Recording Site
Dorsum of the ankle, A is placed between the tibialis anterior tendon and the lateral malleolus. R is 3.5 cm distal to A. G is between the stimulating and recording sites (Fig. 5.12).

Stimulation Site
Lateral calf 14 cm proximal to A (Fig. 5.12).

Control Values
35 subjects [9].
    Latency (ms): 2.7, range 2.1–3.4.
    Amplitude (µV): 13.8, range 0–45.
    CV (m/s): 51.8, range 41–67 (Fig. 5.13).

5.6 Superficial Peroneal Sensory Study

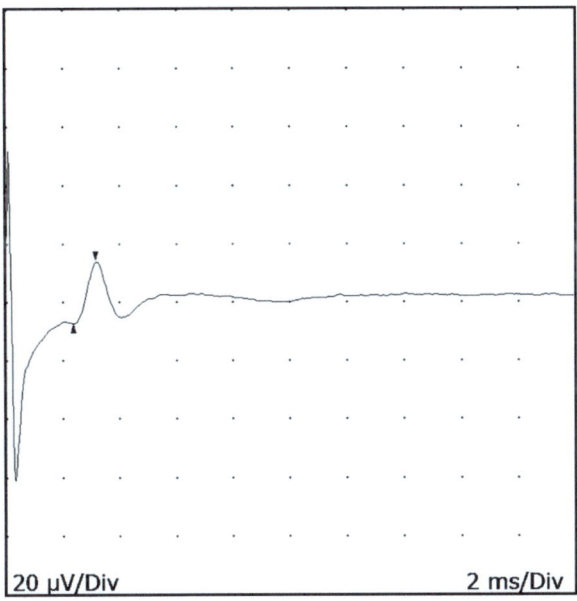

Fig. 5.13 Superficial peroneal sensory study, control subject. SNAP latency is 2.4 ms, amplitude is 18 µV, CV is 48 m/s

Comments

- The nerve branches on the dorsum of the foot. The medial branch passes just lateral to the tendon of the extensor hallucis longus. The intermediate branches lie 1–2 cm medial to the lateral malleolus. These branches are superficial and can be palpated, especially when the foot is plantarflexed and inverted.
- To maximize the SNAP amplitude, the recording electrodes may have to be repositioned either slightly medially or laterally.
- Of 8.6% normal subjects, at least one had no recordable superficial peroneal nerve SNAP.
- Although the normal value for the latency is based on the distance of 14 cm, try a shorter distance if high currents (more than 25 mA) are needed to obtain the response.
- Side-to-side comparisons of the amplitude and latency can be helpful.
- The test is especially useful for differential diagnosis of lesions of the peroneal or sciatic nerves, lumbosacral plexopathy (abnormal SNAP) with L5 radiculopathy proximal to the sensory ganglion (normal SNAP).

5.7 Tibial Nerve Study

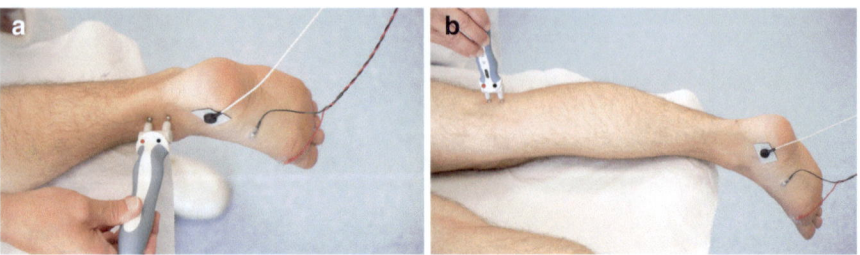

Fig. 5.14 Tibial nerve study. (**a**) stimulation at the ankle, (**b**) stimulation at the popliteal fossa

Nerve Fibers Tested and Route S1 and S2 roots, anterior division of the lumbosacral plexus, sciatic nerve, tibial nerve.

Recording Site
Abductor hallucis brevis (AHB) muscle, A is placed 1 cm proximal and 1 cm inferior to the navicular prominence (the most superior point of the arch formed by the junction of the plantar skin and dorsal foot skin), and R is placed over the medial surface of the metatarsophalangeal joint of the great toe. G is between the stimulating and recording sites (Fig. 5.14).

Stimulation Sites
At the ankle (A) with the cathode slightly proximal and posterior to the medial malleolus 8 cm proximal to the recording electrode; proximal stimulation at mid-popliteal fossa (PF). The anode is proximal at both the stimulation sites (Fig. 5.14).

Normal Values
250 subjects [10].
  Latency (ms): 4.5 ± 0.8, mean + 2 SD 6.1, 97th percentile 6.1.
  Amplitude (mV): 12.9 ± 4.8, mean-2 SD 3.3 mV, 3rd percentile 4.4.
  CV (m/s): 47 ± 6, mean-2 SD 35, 3rd percentile 39 (Fig. 5.15).

5.7 Tibial Nerve Study

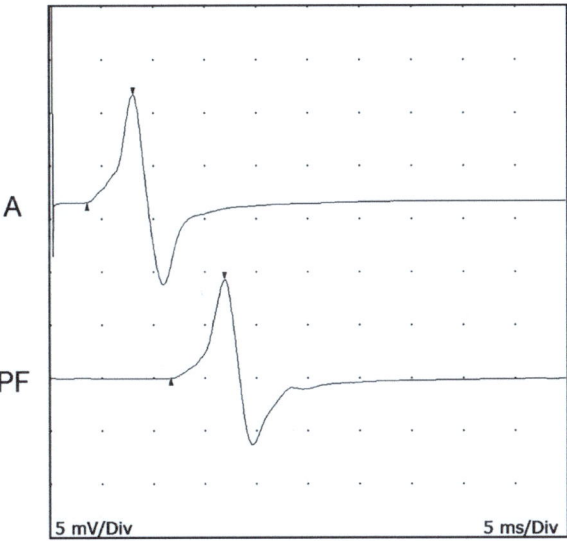

Fig. 5.15 Tibial nerve study, control subject, stimulation at ankle (A) and popliteal fossa (PF): DML is 3.6 ms, distal CMAP amplitude is 10.2 mV, CV is 49.4 m/s

Comments

- The tibial CMAP may present an initial positive deflection, indicating that the recording electrode is not over the motor end plate of the abductor hallucis brevis muscle or records a volume-conducted CMAP generated by other tibial-innervated muscles in the foot. If this occurs, the position of the recording electrode should be changed slightly, but it is not always possible to eliminate the initial positivity. It is debated where to measure the onset latency of a CMAP with a positive onset, and independently if it is measured at the onset or at the peak of the positive deflection, it should be kept in mind that the measurement of the distal latency is not completely reliable for assessing a distal nerve pathology. However, in the case of CMAPs with a positive onset, CV can be calculated paying attention to the fact that the onset latency is measured at the same point in CMAPs from distal and proximal stimulation sites.

- High intensities and long-duration stimulations often are required at the popliteal fossa to ensure supramaximal stimulation (Fig. 5.16).

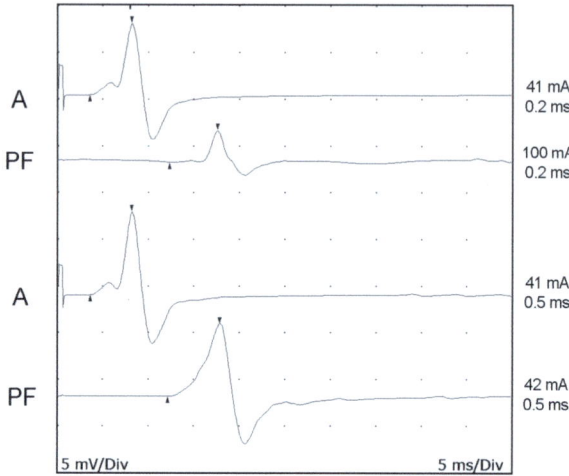

Fig. 5.16 Tibial nerve study, overweight control subject, stimulation at ankle (A) and popliteal fossa (PF). Even using maximal intensity stimulation (100 mA) the amplitude of CMAP from PF stimulation is only 45% of CMAP from A stimulation. Increasing stimulus duration from 0.2 ms to 0.5 ms produces CMAPs of similar amplitude

- The CMAP amplitude at the popliteal fossa stimulation often is lower than at the ankle (in normal subjects may drop up to 50%). Thus, caution must be used whenever interpreting a drop in amplitude between the ankle and popliteal fossa as a conduction block. Side-to-side comparisons often are useful in this situation.
- If the subject has long feet, the fixed 8 cm distance between stimulating and recording electrodes may not include the entire tarsal tunnel. In this case, a 10 cm distance is recommended.

5.8 Medial and Lateral Plantar Nerve Motor Studies

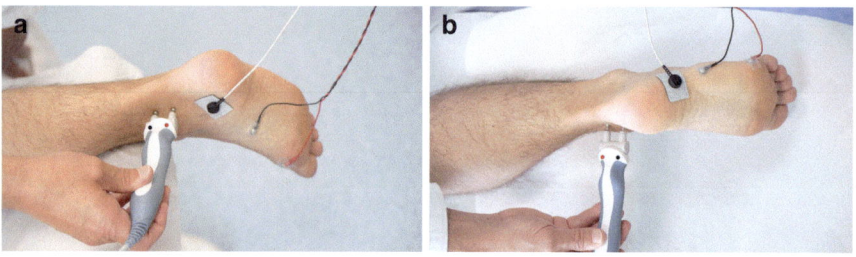

Fig. 5.17 Medial (**a**) and lateral (**b**) plantar nerves motor studies

5.8 Medial and Lateral Plantar Nerve Motor Studies

Nerve Fibers Tested and Route S1 and S2 roots, anterior division of the lumbosacral plexus, sciatic nerve, tibial nerve, medial and lateral plantar nerves.

Recording Sites
Medial plantar: abductor hallucis brevis (AHB) muscle. A is placed 1 cm proximal and 1 cm inferior to the navicular prominence (the most superior point of the arch formed by the junction of the plantar skin and dorsal foot skin), and R is placed over the medial surface of the metatarsophalangeal joint of the great toe (Fig. 5.17a). Lateral plantar, abductor digiti quinti pedis (ADQP) muscle: on the lateral foot, A is placed directly below the lateral malleolus halfway between the tip of the malleolus and the sole of the foot. R is placed over the metatarsophalangeal joint of the little toe (Fig. 5.17b). G is placed between the stimulating and recording sites.

Stimulation Site
The cathode is slightly proximal and posterior to the medial malleolus, and the anode is proximal. The distance is 8–10 cm for AHB, and for ADQP, the distance measurement with an obstetric caliper is required (Fig. 5.17).

Control Values
37 subjects [11].
   Medial plantar latency (ms).
   Distance 8 cm: 3.4 ± 0.5, mean + 2 SD 4.4.
   Distance 10 cm: 3.8 ± 0.5, mean + 2 SD 4.8.
   Lateral plantar latency (ms).
   Distance 8 cm: 3.6 ± 0.5, mean + 2 SD 4.6.
   Distance 10 cm: 3.9 ± 0.5, mean + 2 SD 4.9 (Figs. 5.18 and 5.19).

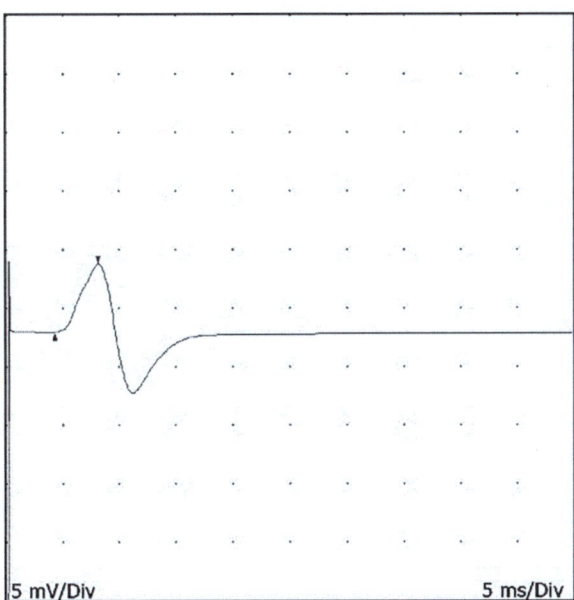

Fig. 5.18 Medial plantar nerve motor study, control subject (distance 10 cm): DML is 4.3 ms, CMAP amplitude is 5.8 mV

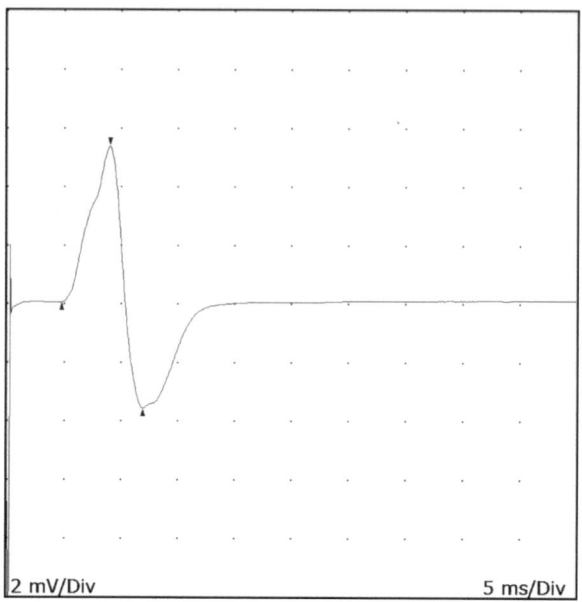

Fig. 5.19 Lateral plantar nerve motor study, control subject (distance 10 cm): DML is 4.8 ms, CMAP amplitude is 5.3 mV

Comments

- CMAP of AHB or ADQP muscles often has an initial positive deflection, indicating that A is not over the motor end plate of the muscle or records far-field potentials generated by other tibial-innervated muscles in the foot. If this occurs, the position of A should be changed slightly, but it is not always possible to eliminate the initial positivity of CMAP.
- Side-to-side comparisons of the amplitude and latency are recommended.
- This study is useful in the evaluation of distal tibial neuropathy across the ankle (i.e., tarsal tunnel syndrome).

5.9 Medial and Lateral Plantar Nerve Sensory Studies

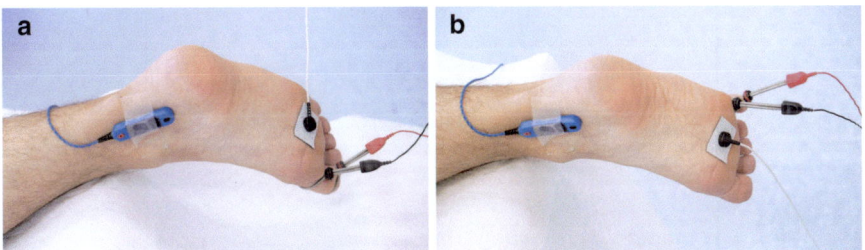

Fig. 5.20 Medial (**a**) and lateral (**b**) plantar nerve sensory studies

Nerve Fibers Tested and Route Medial and lateral plantar nerves, tibial nerve, sciatic nerve, anterior division of the lumbosacral plexus, S1 and S2 roots.

Recording Site
By a bar electrode, A is placed slightly proximal and posterior to the medial malleolus above the flexor retinaculum, and R is 3 cm proximal. G is placed between the stimulating and recording sites (Fig. 5.20).

Stimulation Sites
By ring electrodes. Medial plantar: on the great toe with the cathode placed proximally at the metatarsophalangeal joint of the great toe; the anode placed 3–4 cm distally (Fig. 5.20a). Lateral plantar: on the little toe with the cathode placed proximally at the metatarsophalangeal joint; the anode placed as distally as possible (Fig. 5.20b). Distance: variable, the lateral plantar nerve distance measurement with an obstetric caliper is required.

Control Values
20 subjects [12].
Medial plantar:
Mean amplitude (µV): 3.61, range 2–6.
CV(m/s): 35.22 ± 3.63.
Lateral plantar:
Mean amplitude (µV): 1.89, range 1–5.
CV (m/s): 31.68 ± 4.39.

In this technique, the latency to calculate CV is measured at the negative peak of SNAP; therefore, CV does not correspond to maximal conduction velocity (Figs. 5.21 and 5.22).

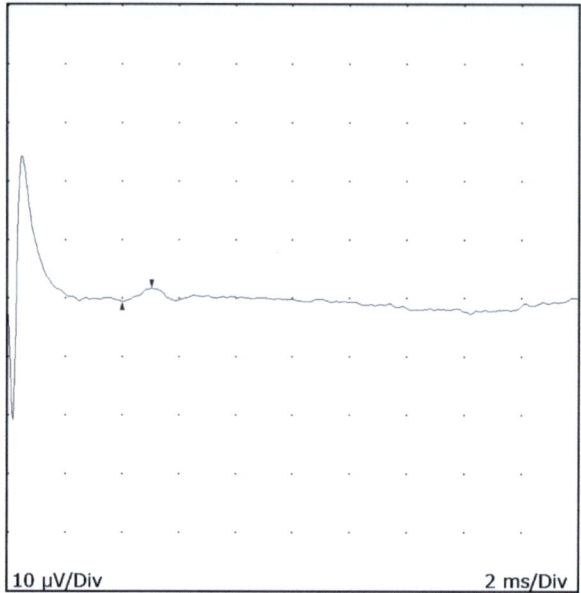

Fig. 5.21 Medial plantar nerve sensory study, control subject: SNAP amplitude is 2.1 µV, CV is 36.2 m/s

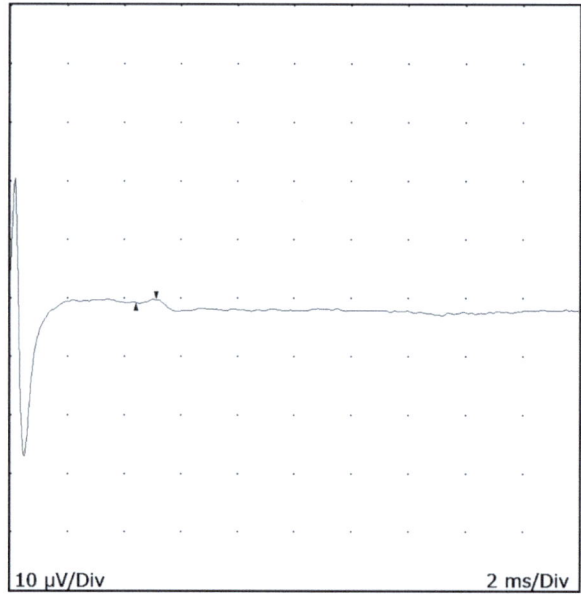

Fig. 5.22 Lateral plantar nerve sensory study, control subject: SNAP amplitude is 2.5 µV, CV is 34.5 m/s

Comments

- The SNAP amplitudes are very small and difficult to obtain, even in normal controls, and averaging is required.
- Side-to-side comparisons of the amplitude and latency are required.
- Side-to-side comparison is necessary before interpreting a low amplitude or absent potential as abnormal.
- This study is useful in the evaluation of distal tibial neuropathy across the ankle (i.e., tarsal tunnel syndrome).

5.10 Sural Nerve Study

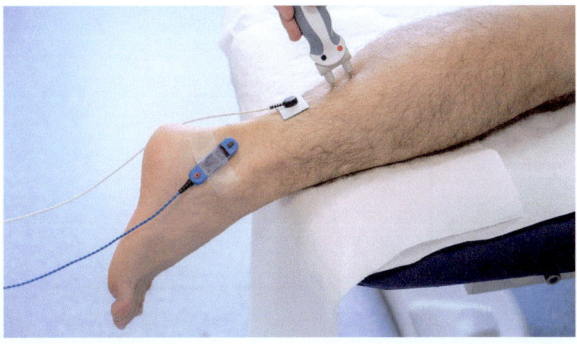

Fig. 5.23 Sural nerve study

Nerve Fibers Tested and Route
Sural nerve, tibial nerve, and by a communicating branch to the peroneal nerve, sciatic nerve, anterior and posterior division of lumbosacral plexus, S1 and S2 roots.

Recording Site
By a bar electrode, A is placed posterior to the lateral malleolus, and R is 3 cm distally. G is between the stimulating and recording sites (Fig. 5.23).

Stimulation Site
The posterior-lateral calf cathode at a distance of 14 cm from A. The anode is proximal (Fig. 5.23).

Control Values
80 subjects. The SNAP is not recordable in two subjects [13].
  Latency (ms): 2.9 ± 0.3, range 2.3–3.7.
  Peak-to-peak amplitude (µV): 16.6 ±, range 5–56 (Fig. 5.24).
  122 elderly subjects, mean 74.1 years, range 60–89 years. The SNAP is not recordable in four subjects [14].

Latency (ms), 3.2 ± 0.4; mean + 2SD, 4.0.
Onset-to-peak amplitude (µV): 10.3 ± 5.8.

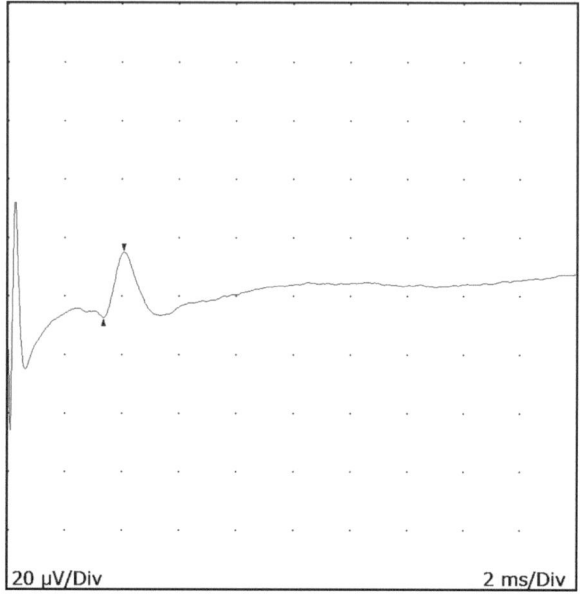

Fig. 5.24 Sural nerve study, control subject. SNAP latency is 3.4 ms, SNAP amplitude is 23.2 µV

Comments

- Supramaximal stimulation usually can be achieved with low stimulation intensities (e.g., 5–25 mA). To maximize the response, the recording electrodes may have to be repositioned either slightly medially or laterally.
- If the SNAP is not recordable stimulating at 14 cm, try a shorter distance of 10–12 cm. If a good response is obtained, do not use the onset latency to determine if the response is normal, but calculate conduction velocity based on the distance used. CV is generally considered normal when ≥40 m/s.
- The SNAP may be unobtainable in healthy elderly subjects.
- The SNAP may be abnormal in lesions of the sciatic nerve, lumbosacral plexopathies, and polyneuropathies; it is normal in S1 radiculopathies proximal to the sensory ganglion.
- The sural SNAP amplitude may be preserved, when compared to the upper extremity nerve SNAPs, in the early phase of Guillain-Barré syndrome.

References

Femoral Nerve Study

1. Stöhr M, Schumm F, Ballier R. Normal sensory conduction in the saphenous nerve in man. Electroencephalogr Clin Neurophysiol. 1978;44:172–8.

Lateral Femoral Cutaneous Nerve Study

2. Laroy V, Knoops P, Semoulin P. The lateral femoral cutaneous nerve: nerve conduction technique. J Clin Neurophysiol. 1999;16:161–3.

Saphenous Nerve Study

3. Wainapel SF, Kim DJ, Ebel A. Conduction studies of the saphenous nerve in healthy subjects. Arch Phys Med Rehabil. 1978;59:316–9.

Peroneal Nerve Study Recording from the Extensor Digitorum Brevis Muscle

4. Kimura J. Electrodiagnosis in diseases of nerve and muscle: principles and practice. 4th ed. OUP USA; 2013.
5. Uncini A, Di Muzio A, Awad J, Gambi D. Compressive bilateral peroneal neuropathy: serial electrophysiologic studies and pathophysiological remarks. Acta Neurol Scand. 1992;85:66–70.

Peroneal Nerve Study Recording from the Tibialis Anterior Muscle

6. Devi S, Lovelace RE, Duarte N. Proximal peroneal nerve conduction velocity: recording from anterior tibial and peroneus brevis muscles. Ann Neurol. 1977;2:116–9.
7. DeLisa JA, MacKenzie K, Baran M. Manual of nerve conduction velocity and somatosensory evoked potential. 2nd ed. New York: Raven Press; 1987.
8. Escorcio-Bezerra ML, Abrahao A, Nunes KF, de Castro Sparapani FV, de Oliveira Braga NI, Robinson LR, Zinman L, Manzano GM. Optimal E2 (reference) electrode placement in fibular motor nerve conduction studies recording from the tibialis anterior muscle. Muscle Nerve. 2019;59:249–53.

Superficial Peroneal Sensory Study

9. Levin KH, Stevens JC, Daube JR. Superficial peroneal nerve conduction studies for electromyographic diagnosis. Muscle Nerve. 1986;9:322–6.

Tibial Nerve Study

10. Buschbacher RM. Tibial nerve motor conduction to the abductor hallucis brevis. Am J Phys Med Rehabil. 1999;78:S15–20.

Medial and Lateral Plantar Nerve Motor Studies

11. Fu R, DeLisa JA, Kraft GH. Motor nerve latencies through the tarsal tunnel in normal adult subjects. Arch Phys Med Rehabil. 1980;51:164–9.

Medial and Lateral Plantar Nerve Sensory Studies

12. Oh SJ, Sarala PK, Kuba T, Elmore RS. Tarsal tunnel syndrome: electrophysiological study. Ann Neurol. 1979;5:327–30.

Sural Nerve Study

13. Izzo LK, Sridhara CR, Rosenholtz H, Lemont H. Sensory conduction studies of the branches of the superficial peroneal nerve. Arch Phys Med Rehabil. 1981;62:24–7.
14. Falco FJE, Hennesey WJ, Goldberg G, Braddon RL. Standardized nerve conduction studies in the lower limb of healthy elderly. Am J Phys Med Rehabil. 1994;73:168–74.

Special Studies

6.1 F Wave Studies

The F wave is a muscle action potential recorded from a small number of motor units or even from a single motor unit, resulting from antidromic activation of the anterior horn cells.

6.1.1 Median and Ulnar Nerves

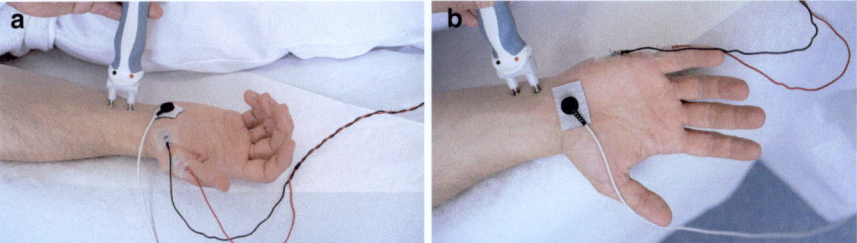

Fig. 6.1 Median (**a**) and ulnar (**b**) F wave studies

Nerve Fibers Tested and Route
Motor nerve fibers activated antidromically, and α-motor neurons and motor nerve fibers activated orthodromically.

Recording Sites
The same as for median and ulnar motor conductions (Fig. 6.1).

Stimulation Sites

The same as for median and ulnar motor conductions. The cathode of the stimulator is proximal to avoid anodal block (Fig. 6.1). The ground electrode (G) is between the stimulating and recording sites or the dorsum of the hand. Stimulation is supramaximal, and at least 16 stimuli must be delivered.

To calculate F wave conduction velocity, the distance is measured from the point of stimulation to the 7th cervical spine with the subject upright and the arm abducted at 90°. The hand is supinated for the median nerve and pronated for the ulnar nerve. Measurement is made along the courses of the nerves to the axilla and then around the back of the shoulder to the 7th cervical spinous process (just above T1, which is the most prominent process).

F wave conduction velocity is calculated according to the formula:

$$\text{Distance} / \left[\left(\text{F wave minimal latency} \right) - \left(\text{DML} \right) - 1 \right] / 2.$$

Where:

- The distance is measured in mm as described above.
- The F wave minimal latency—DML is the antidromic plus orthodromic F conduction time in ms.
- 1 ms is the estimated time to excite the motor neuron hillock.
- The result is divided by 2 to represent the F conduction time from the stimulation site to the spinal cord.

6.1.2 Peroneal and Tibial Nerves

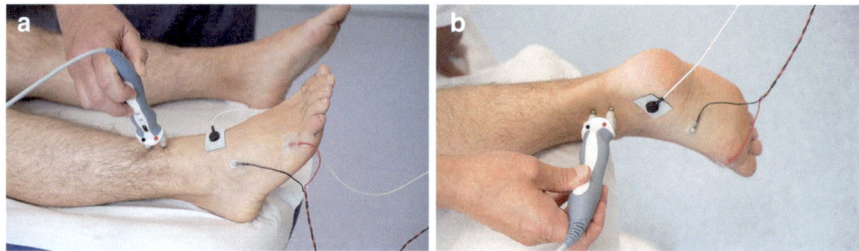

Fig. 6.2 Peroneal (**a**) and tibial (**b**) F wave studies

Nerve Fibers Tested and Route

Motor nerve fibers activated antidromically, and α-motor neurons and motor nerve fibers activated orthodromically.

Recording Sites

The same as for peroneal and tibial motor conduction (Fig. 6.2).

Stimulation Sites

The same as for peroneal and tibial motor conduction. The cathode of the stimulator is proximal to avoid anodal block (Fig. 6.2). G is between the stimulating and recording sites. Stimulation is supramaximal, and at least 16 stimuli must be delivered.

6.1 F Wave Studies

To calculate F wave conduction velocity, the distance for both the peroneal and tibial nerves is measured from the stimulation point to the lower border of the T-12 spinous process by way of the greater trochanter of the femur. F wave conduction velocity is calculated according to the formula reported above.

Control Values

122 median nerves from 61 subjects, 130 ulnar nerves from 61 subjects, 120 peroneal nerves from 60 subjects, 118 tibial nerves from 59 subjects [1].
Minimal F-wave latency (ms), ULN calculated as mean + 2 SD.
Median (wrist), 26.6 ± 2.2; ULN, 31.
Ulnar (wrist), 27.6 ± 2.2; ULN, 32.
Peroneal (ankle), 48.4 ± 4.0; ULN, 56.
Tibial (ankle), 48.4 ± 4.0 ms; ULN, 58.
Minimal F wave latencies show a linear correlation with the subject's height [2] (Fig. 6.3).

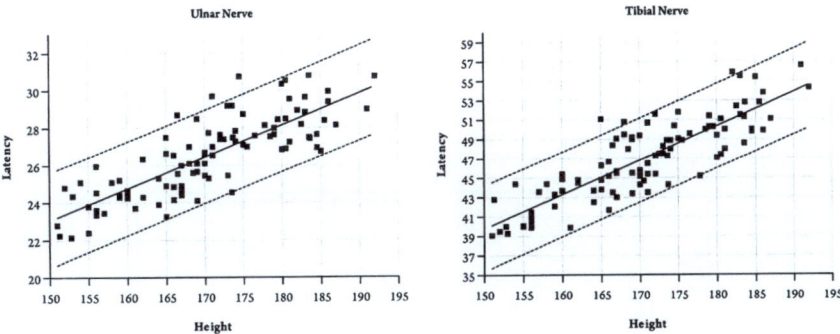

Fig. 6.3 Nomograms of height-minimal F wave latency for ulnar and tibial nerves. Mean (solid line) and ULN and LLN (dotted lines) calculated as mean ± 2.5 SD are shown. (Reproduced from Nobrega et al. [2] with permission by Elsevier)

Side-to-side minimal F wave latency difference (ms), ULN calculated as mean + 2 SD.
Median (wrist), 0.93 ± 0.62; ULN, 2.2.
Ulnar (wrist), 0.85 ± 0.58; ULN, 2.0.
Peroneal (ankle), 0.76 ± 0.52; ULN, 1.8.
Tibial (ankle), 1.52 ± 1.02; ULN, 3.6.

F wave conduction velocity (m/s), LLN calculated as mean − 2 SD.
Median (wrist), 65.3 ± 4.7; LLN, 56.
Ulnar (wrist), 65.3 ± 4.78; LLN, 55.
Peroneal (ankle), 49.8 ± 3.6; LLN, 43.
Tibial (ankle), 52.6 ± 4.3; LLN, 44.
In Figs. 6.4 and 6.5 are reported examples of F wave recordings from median and tibial nerve stimulation.

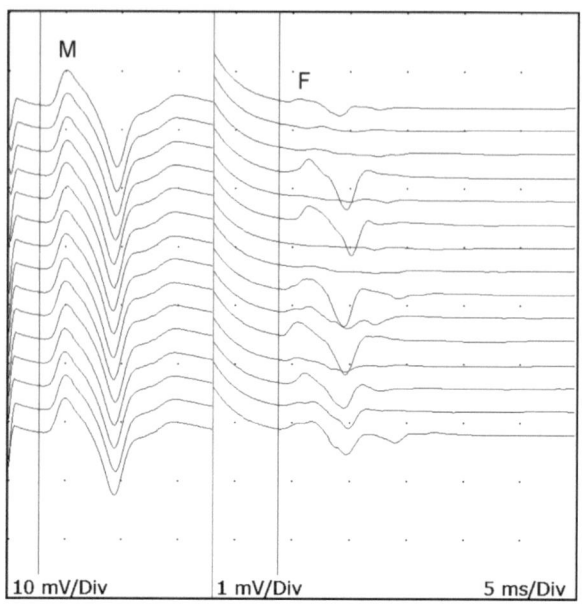

Fig. 6.4 Median nerve, control subject: M wave (M) and F wave (F). Minimal F wave latency is 23.8 ms

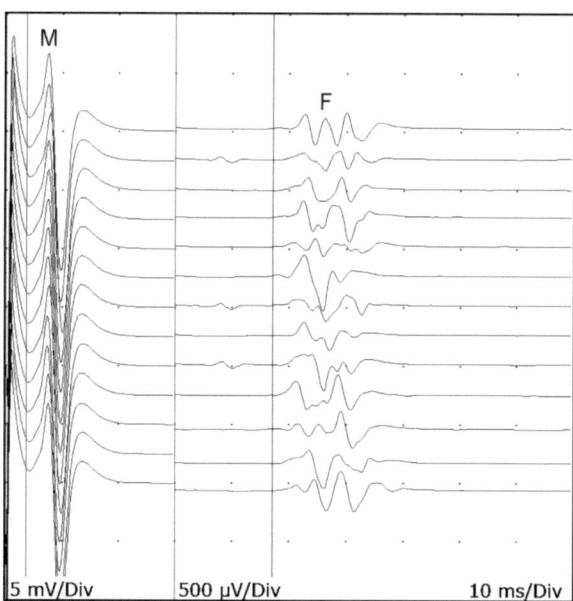

Fig. 6.5 Tibial nerve, control subject: M wave (M) and F wave (F). Minimal F wave latency is 47.2 ms. Note the high persistence of F (100%)

Comments
- The F wave is a recurrent motor response, not a reflex. It results from the backfiring of antidromically activated α-motor neurons.
- In proximal short nerves, the F wave is difficult to detect because it is close to the end of the M wave.
- The F wave latency decreases when the point of stimulation is moved proximally.
- The F wave is not always recordable with each stimulus.
- In the peroneal nerve, persistence of the F wave (number of F waves/number of stimuli delivered) is low; the F wave cannot even be recorded in normal subjects.
- In the tibial nerve, F wave persistence is high (usually 100%).
- The F wave is present only if a M wave is elicited.
- The F wave may be absent when the M wave amplitude is <1 mV.
- The F wave is variable in latency and morphology.
- For tall or short subjects, the F wave minimal latency should be normalized for height.
- The F wave receives innervation from multiple roots; therefore, it may be normal when only one root is injured.
- A prolonged F wave latency, in the presence of normal distal and intermediate nerve segment conductions, suggests a proximal neuropathy, plexopathy, or radiculopathy. However, this finding cannot be used to differentiate these conditions.

6.2 Soleus H Reflex Study

The H reflex is an electrically elicited spinal monosynaptic reflex.

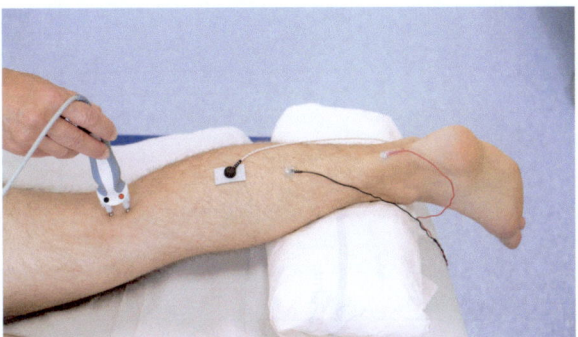

Fig. 6.6 Soleus H reflex study

Nerve Fibers Tested and Route
Afferent large-diameter fibers (Ia) of the tibial nerve, S1 root, spinal cord, efferent fibers of the tibial nerve.

Recording Site
With the subject prone, the lower leg over a pillow, and the foot hanging over the edge of the bed. A is placed one to two fingerbreadths distal to where the soleus meets the two bellies of the gastrocnemius; R is placed over the Achilles tendon. Ground is placed between the stimulating and recording sites (Fig. 6.6).

Stimulation Site The cathode is placed at the mid-popliteal fossa, and the anode is distal. Distance to A is variable: usually 20–25 cm (Fig. 6.6).

Stimulation Features stimulus duration is 0.5–1 ms to more selectively activate the Ia sensory fibers. H reflex occurs with low stimulation intensities. The stimuli should be applied randomly every few seconds in order to avoid habituation (stimulation rate <0.5 Hz) (see Course lesson "Late responses").

Control Values
251 adult subjects, height (cm): <160–≥180 [3].
  Latency (ms), 30.3 ± 2.4; ULN (mean + 2 SD), 35.1; 97th percentile, 35.0 (Fig. 6.7).

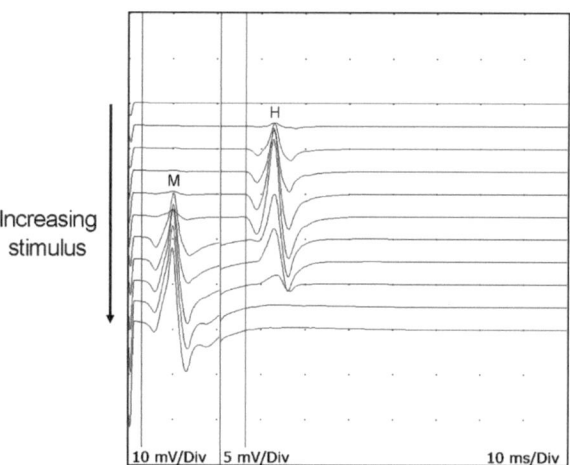

Fig. 6.7 Serial M wave (M) and H reflex (H) recording in a control subject. As the stimulus is slowly increased, the H reflex appears first without the M wave; as the current is increased, the H reflex increases in amplitude and a M wave is also recorded; as the current is further increased, M wave amplitude increases, whereas the H reflex decreases, till it disappears. M latency is 3.0 ms, and H latency is 26.5 ms

Comments

- The H reflex is a monosynaptic reflex recorded from the soleus muscle and elicited by electrical stimulation of its muscle spindle afferents (Ia fibers).
- This electrically elicited reflex is similar to the mechanically induced tendon reflex. The main difference is that the H reflex bypasses the muscle spindle and the Ia fibers are synchronously stimulated. This can explain why the ankle jerk can be absent and the H reflex be still recordable. The presence of the ankle jerk with a non-recordable H reflex indicates some technical problem.
- The distance between stimulating and recording electrodes must be the same on both sides to ensure a correct side-to-side comparison.
- Stimulus duration must be set at 0.5–1 ms to more selectively activate the Ia sensory fibers.
- H reflex occurs with low stimulation intensities.
- Increasing stimulation intensity, the M response is evoked and the H amplitude diminishes, because there is a collision in the motor fibers between the antidromic potentials and the reflex potentials.
- The stimuli should be applied randomly every few seconds in order to avoid habituation.
- H reflex usually has a triphasic morphology (positive, negative, positive) and a latency of 25–36 ms.
- For tall or short subjects, H reflex latency should be normalized for height.
- Comparison to the contralateral side is often helpful in determining if a latency is abnormal (latency difference >2.0 ms) [3].
- A side-to-side peak-to-peak amplitude ratio smaller than 0.4 is considered abnormal [4].
- The H reflex (as the ankle jerk) is absent in many healthy subjects over the age of 70 years.
- H reflex is delayed or absent in polyneuropathies, tibial neuropathy, sciatic neuropathy, lumbosacral plexopathy, or S1 radiculopathy.

6.3 Bulbocavernosus Reflex Study

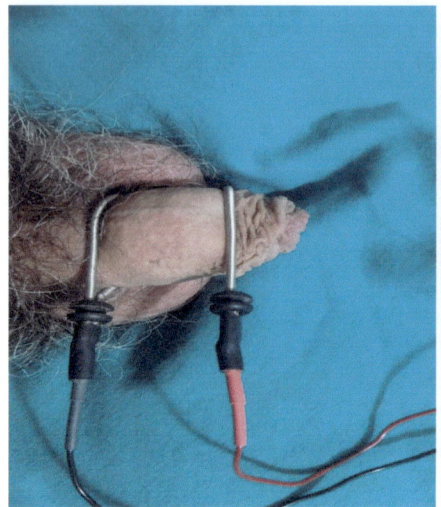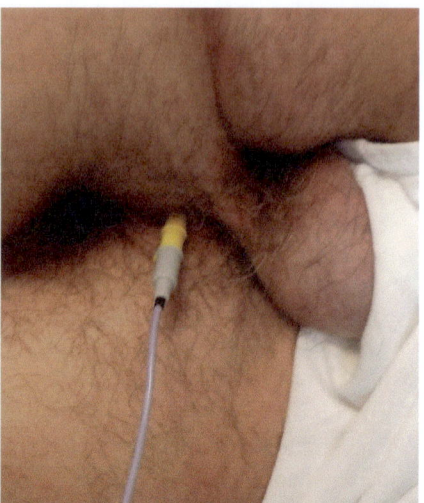

Fig. 6.8 Bulbocavernosus reflex study

Nerve Fibers Tested and Route
Afferent branch: dorsal nerve of penis (pudendal nerve), spinal cord (S2–S4). Efferent branch: lower hemorrhoid nerve (pudendal nerve).

Recording Site
With a concentric needle electrode inserted in the bulbocavernosus muscle just behind the scrotum, G is positioned on the thigh (Fig. 6.8).

Stimulation Site
By using ring electrodes of the dorsal nerve of the penis, the cathode is proximal and the anode is 2–3 cm distal. The intensity of stimulation required is usually 5–20 mA, and the stimulus duration employed is 0.2–0.5 ms (Fig. 6.8).

The reflex is made by two responses: RI oligosynaptic and R2 polysynaptic (Fig. 6.9).

Control Values
Latency R1 (ms), 35 ± 2.1; range, 28–42 [5].

Several other control value sets have been reported, and it is generally agreed that R1 latency is abnormal when it is >42 ms.

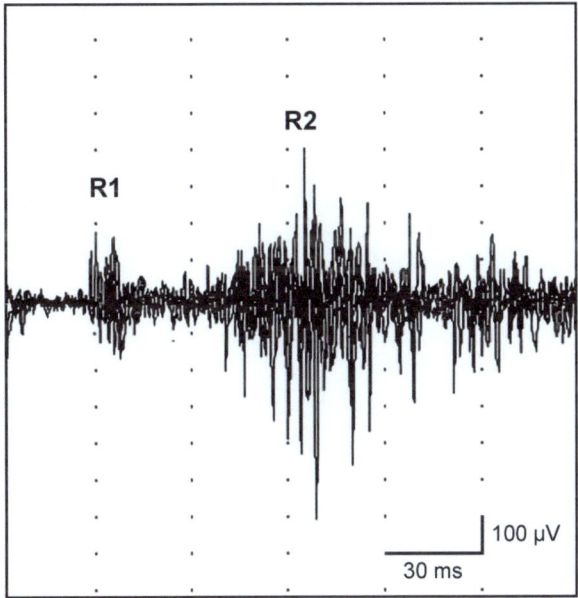

Fig. 6.9 Bulbocavernosus reflex study, control subject: R1 latency is 29 ms

Comments
- The bulbocavernosus muscle is superficial and located under the scrotum on both sides of the midline.
- A correct needle placement can be confirmed before starting the stimulation by squeezing the glans and observing the reflex electric activity in the bulbocavernosus muscle.
- The study can be also performed in women. The clitoris can be stimulated, or a special stimulator can be attached to a Foley catheter and placed intraurethrally.

References

F Wave Studies

1. Kimura J. Electrodiagnosis in diseases of nerve and muscles. 3rd ed. New York: Oxford University Press; 2001. p. 639–465.
2. Nobrega JAM, Pinheiro DS, Manzano G, Kimura J. Various aspects of F-wave values in a healthy population. Clin Neurophysiol. 2004;115:2336–42.

Soleus H Reflex Study

3. Buschbacher RM. Normal range for H-reflex recording from the calf muscles. Am J Phys Med Rehabil. 1999;78:S75–9.
4. Jankus WR, Robinson LR, Little JW. Normal limits of side to side H-reflex amplitude variability. Arch Phys Med Rehabil. 1994;75:3–7.

Bulbocavernosus Reflex Study

5. Siroky MB, Sax DS, Krane RJ. Sacral signal tracing: the electrophysiology of the bulbocavernosus reflex. J Urol. 1979;122:661–4.

Part III

A Guide to Needle Electromyography

Needle Electromyography (EMG)

7

The electromyography (EMG) examination is usually performed after the nerve conduction studies (NCSs) are completed, because the findings of NCSs are useful in the planning and interpretation of the needle examination.

7.1 Basic Principles

The needle electromyography (EMG) is the most difficult part of the electroneuromyography examination. It requires not only the knowledge of anatomy and physiology but also establishing a good relationship with the patient. Although the basics of the needle EMG can be learned usually in a short time, it may take years of experience to recognize the uncommon and subtle EMG findings. The following very basic notions of EMG for the beginner are described.

The EMG consists of recording the electrical activity of the muscle fibers by an intramuscular electrode inserted transcutaneously. The signal is displayed on the screen of the EMG machine and also converted into a sound played by a loudspeaker. To record EMG, concentric or monopolar EMG needles can be used. The concentric needle contains both the active and reference electrodes. The shaft of the needle serves as the reference electrode, whereas the active electrode is a small wire running in the center of the needle and exposed at the tip. The monopolar needle is Teflon coated, and its exposed tip functions as the active recording. It has the advantage of having a smaller diameter and a sharper point and may be slightly less painful and easier to tolerate for the patient. A surface disc electrode is required as the reference electrode and should be placed close to the needle electrode, but this setup produces greater electrical noise compared with the concentric needle. With both needle electrodes, motor unit potentials (MUPs), which are the algebraic sum of single muscle action potentials, can be recorded. There are, however, some differences. With a concentric needle, the MUP amplitude is slightly smaller and the duration is shorter than those obtained with a monopolar needle. However, the normal values for quantitative MUP analysis have been established more definitively in more muscles by concentric needle. Overall, the concentric needle is preferred by most electromyographers. The ground electrode should be applied to the limb being studied for electrical safety reasons.

Before beginning the needle EMG examination, it is important to explain all the steps of the procedure to the patient to alleviate any patient's anxiety and fear. Establishing a good rapport with the patient is essential for the study which cannot be performed without good cooperation.

Disposable gloves should be always worn to prevent the transmission of blood-borne infections between the patient and the electromyographer. For each muscle being studied, the point for needle insertion should be identified, and how to properly activate the muscle should be explained to the patient. Because of the small size of some muscles (as in the hand) or the vicinity and overlap of other muscles (as in the forearm), it is important to know the possible pitfalls and achieve an appropriate technique. After the insertion, the location of the needle is confirmed by asking the subject to activate slightly the muscle of interest. If the needle is properly placed, sharp MUPs with a crisp sound will be recorded. If this is not the case, the needle should be either pulled back slightly or inserted a bit deeper into the muscle. If also this maneuver fails to produce sharp MUPs, the needle must be removed and the insertion point reconsidered.

The EMG activity is evaluated in three different stages: (1) during electrode insertion and at rest, (2) when the muscle is minimally activated, and (3) when the muscle is maximally activated.

7.2 EMG Evaluation During Electrode Insertion and at Rest

This evaluation should be performed with the sensitivity set at 50 μV per division, and the analysis time is set at 20–50 ms per division for assessing insertional activity and at 10–20 ms per division for activity at rest. At least four to six brief needle movements are made in four quadrants of each muscle to assess for insertional activity and spontaneous activity at rest. When the needle is inserted and rapidly moved into the muscle, there is a brief burst of muscle fiber potentials, typically lasting no longer than 300 ms after stopping the needle movement. This burst of electrical activity originates in the muscle fiber mechanically stimulated by the needle electrode. Increased duration of insertional activity may be seen in both neuropathic and myopathic conditions. On the other hand, insertional activity may be decreased when the muscle has been replaced by connective and fat tissue. An assessment of the number of fibrotic changes can be roughly made by evaluating the resistance to needle insertion.

The muscle is normally electrically silent at rest, except for the potentials recorded at the end plate zone. Spontaneous activity is usually defined as any activity at rest that lasts longer than 3 seconds. The identification of the many forms of spontaneous activity can be achieved by an analysis of the waveform, stability, firing characteristics, and sound. The type and degree of abnormal electrical activity at rest (a) suggest the neuroanatomic source (muscle fiber or motor unit), (b) can provide diagnostic information (i.e., myotonic discharges are seen only in myopathies), (c) may help to determine the severity of the lesion, and (d) can suggest the time course of the injury. For a detailed analysis and identification of the types of spontaneous activity, the reader is referred to classical textbooks of electroneuromyography [1, 2].

7.3 EMG Evaluation at Minimal Voluntary Effort

Once insertional and spontaneous activity has been studied, with the needle left in place, the characteristics of individual MUPs are evaluated. The analysis time is usually set at 10 ms per division, the sensitivity is set at 200 µV per division to asses MUP duration, and it varies for assessing the MUP amplitude. To analyze MUPs, the examiner asks the patient to minimally contract the muscle recruiting few clearly identifiable MUPs. Then the needle is gradually moved until the MUP of interest becomes "sharp" (sounds louder and crisper). The closer the needle to the MUP, the higher the amplitude and the shorter the major spike rise time. This procedure allows the electromyographer to study the duration, amplitude, and morphology (phases) of MUPs. The MUP duration is the most reliable parameter to assess; it varies according to the muscles examined and the age of the subject (Fig. 7.1).

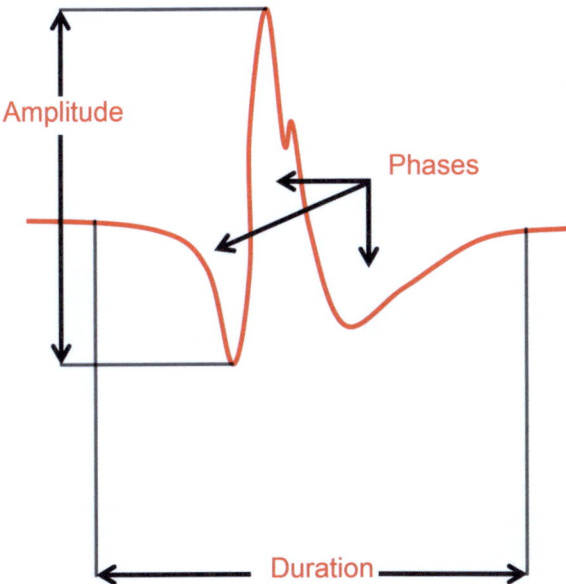

Fig. 7.1 MUP parameters

After one or two MUPs are assessed at one location, the needle is moved slightly within the muscle to a different site, and the process is repeated. For visual motor unit analysis, at least ten different MUPs should be sampled. In primary muscle disease, the individual muscle fibers degenerate producing short-duration, low-amplitude MUPs. In anterior horn cell diseases or axonal neuropathies, the muscle fibers of the motor unit are denervated. Because of collateral sprouting, the remaining motor units, located in the same area, may partially reinnervate the denervated muscle fibers resulting in MUPs with increased amplitude, duration, and number of phases (polyphasic MUP).

For an exhaustive explanation of MUP parameters and the pathological correlations, the reader is referred to the classical textbooks of electroneuromyography [1, 2].

7.4 EMG Evaluation at Maximal Voluntary Effort

After evaluating single MUPs, the patient is asked to make a maximal muscle contraction. The analysis time is usually set at 200 ms per division, and the sensitivity is set at 1 mV per division. In the normal subject, the number of MUPs firing is so increased that the details of individual MUPs are lost and the baseline cannot be distinguished any longer. This is called a "full interference pattern." In pathological conditions, the interference pattern changes. In myopathies, the number of motor units is unchanged, but the number of muscle fibers in the motor units is decreased. Because each motor unit generates less force, the patient will have to recruit a large number of motor units to generate a small amount of strength. This phenomenon is called "early recruitment" and is one of the key points in establishing the diagnosis of a "myopathic pattern." In anterior horn cell diseases and peripheral neuropathies, the number of motor units is reduced, producing an incomplete interference pattern of MUPs firing at high frequency.

The amplitude of the interference pattern can be measured by drawing parallel lines through the maximal positive and negative peaks, excluding solitary high peaks (envelope). The normal value for the envelope amplitude differs among muscles but is generally <4 mV.

7.5 Steps in Needle EMG

The steps in needle EMG examination are summarized as follows:

- Explain the procedure to the patient.
- Wear disposable gloves.
- Select the muscle to study.
- Show the patient how to activate the muscle.
- Palpate the muscle during contraction.
- Locate the needle insertion point in the muscle.
- Ask the patient to relax the muscle.
- Clean the skin with alcohol.

- Insert the needle into the relaxed muscle.
- Ask the patient to contract the muscle slightly to ensure proper placement.
- To assess insertional and spontaneous activity, ask the patient to relax the muscle completely.
- Make short movements of the needle in all four quadrants.
- To assess individual MUPs, ask the patient to contract the muscle slightly, and gently move the needle until the MUP becomes "sharp." Assess several locations for the MUP duration, amplitude, and number of phases.
- Assess MUP recruitment at maximal effort, and measure the envelope.

7.6 Anatomy for Electromyography

In the following chapters, the muscles commonly tested at EMG are grouped according to the nerve supply and reported in a cranial-caudal order. Diagrams of plexuses and peripheral nerves with the muscles that they supply are presented. For each muscle, the innervation and the nerve fibers route are described. Instructions are provided on how to test the strength and how to activate it during the EMG examination.

The needle insertion point is shown and described using, when appropriate, anatomical landmarks. By means of anatomical cross sections, oriented according to the perspective of the examiner (the eye in the figure), the relationships among the muscle to be examined and the adjacent muscles/structures are demonstrated.[1] Moreover, it is remarked how to avoid common pitfalls in EMG examination.

Selected References

1. Kimura J. Electrodiagnosis in diseases of nerves and muscles: principles and practice. 3rd ed. New York: Oxford University Press; 2001.
2. Preston DC, Shapiro BE. Electromyography and neuromuscular disorders. 3rd ed. London: Elsevier Saunders; 2013.

---

[1] All the cross sections are from: A Cross-Section Anatomy by Albert C. Eycleshymer and Daniel M. Schoemaker, D Appleton Century Company, New York, 1911. The figures are in the public domain and have been modified and oriented according to the examiner's viewpoint. Anatomic terms are according to the original in the Basle Nomina Anatomica. In these figures: A – artery, N = nerve, V = vein.

Muscles Innervated by the Cranial Nerves

8.1 Trigeminal Nerve, Masseter

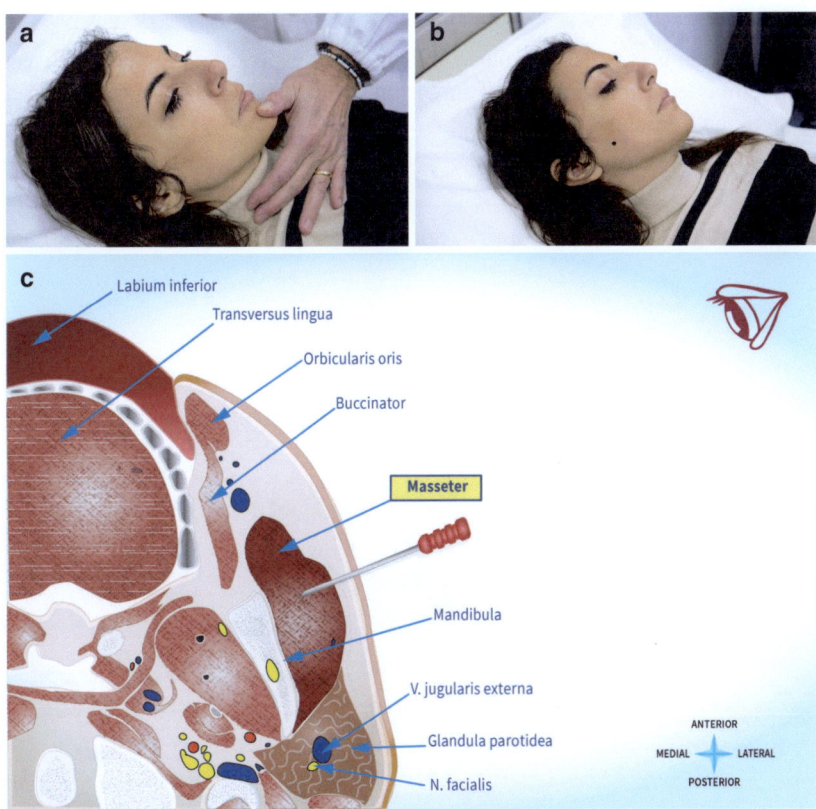

Fig. 8.1 Masseter: (**a**) test, (**b**) insertion point, (**c**) cross-section through the lower portion of the mandibular ramus. The eye shows the perspective of the examiner

Innervation and Nerve Fibers Route
Trigeminal motor nucleus (pons), lower trigeminal branch (V3), mandibular nerve.

Test and Activation
Ask the subject to clench the jaw. Apply pressure to the chin in the direction of opening the mouth (Fig. 8.1a).

Needle Insertion
Palpate the muscle as the patient clenches the jaw, and insert the needle two fingerbreadths anterior to the angle of the mandible and one to two fingerbreadths superior to the lower edge of the mandible (Fig. 8.1b).

Caveat: if the needle is inserted too posteriorly, it will go through the parotid gland and eventually damage the facial nerve and the external jugular vein; if inserted too cephalad, close to the zygomatic arch, it may damage the main parotid duct (Fig. 8.1c).

Comments
- The masseter is useful to sample in patients with suspected motor neuron disease.
- In the masseter, the duration of motor unit action potential is normally briefer than in the limb muscles.

8.2 Facial Nerve

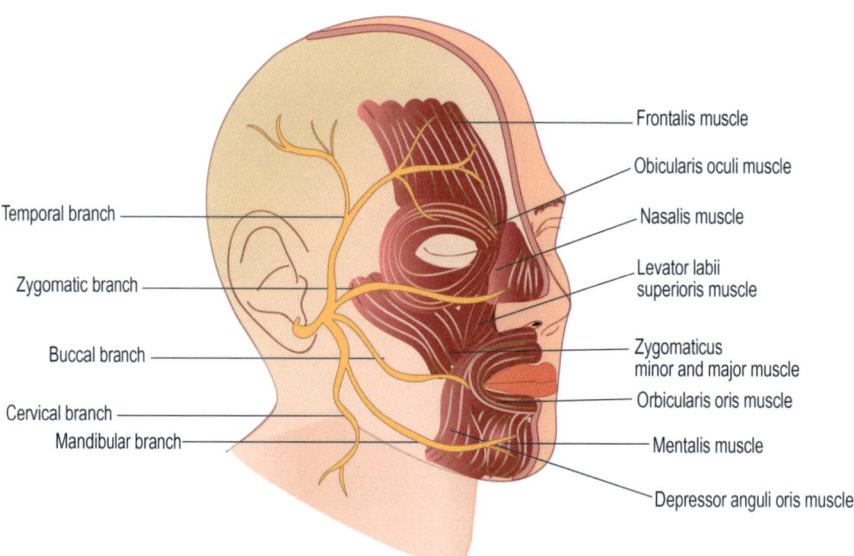

Fig. 8.2 The facial nerve, its major branches, and supplied muscles

8.2.1 Frontalis

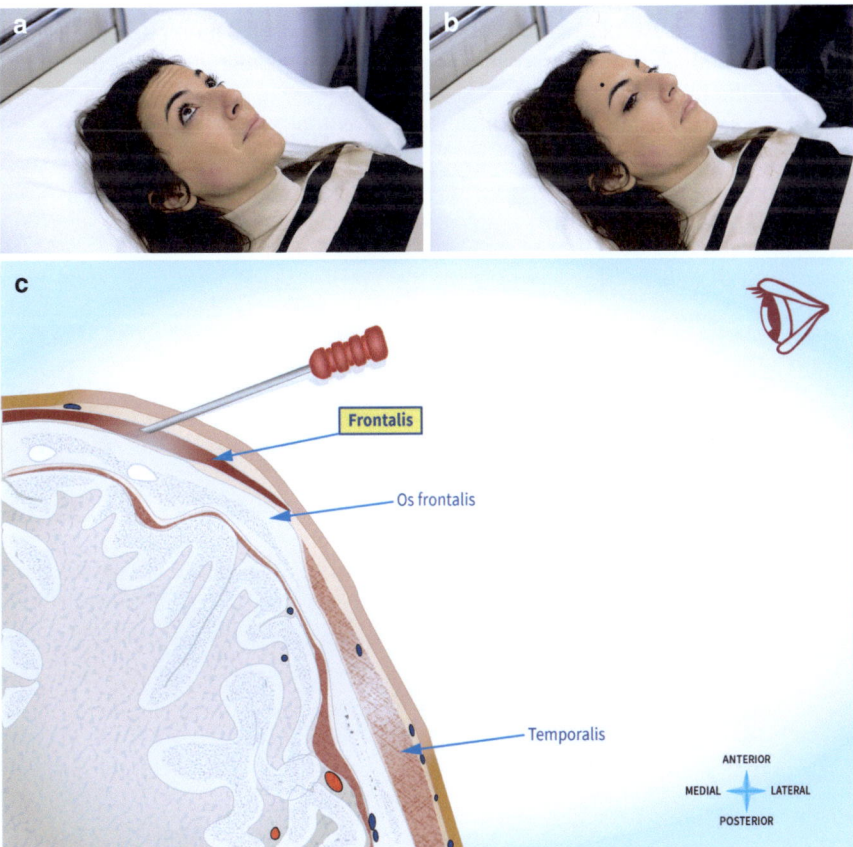

Fig. 8.3 Frontalis: (**a**) test, (**b**) insertion point, (**c**) cross section through the upper margin of the temporal bone. The eye shows the perspective of the examiner

Innervation and Nerve Fibers Route
Facial nucleus (pons), facial nerve, temporal branch.

Test and Activation
Ask the subject to look up wrinkling the forehead (Fig. 8.3a).

Needle Insertion
With the subject lying supine and the head in a neutral position, insert the needle tangentially, one/two fingerbreadths above the middle of the eyebrow (Fig. 8.3b).
 Caveat: the muscle is very thin (Fig. 8.3c).

Comments

- The frontalis is tested in assessing Bell's palsy.
- The frontalis is useful to sample in patients with suspected motor neuron disease.
- In the frontalis, the duration of the motor unit action potential is normally briefer than in the limb muscles.
- The frontalis is used for single-fiber electromyography (EMG).

8.2.2 Orbicularis Oculi

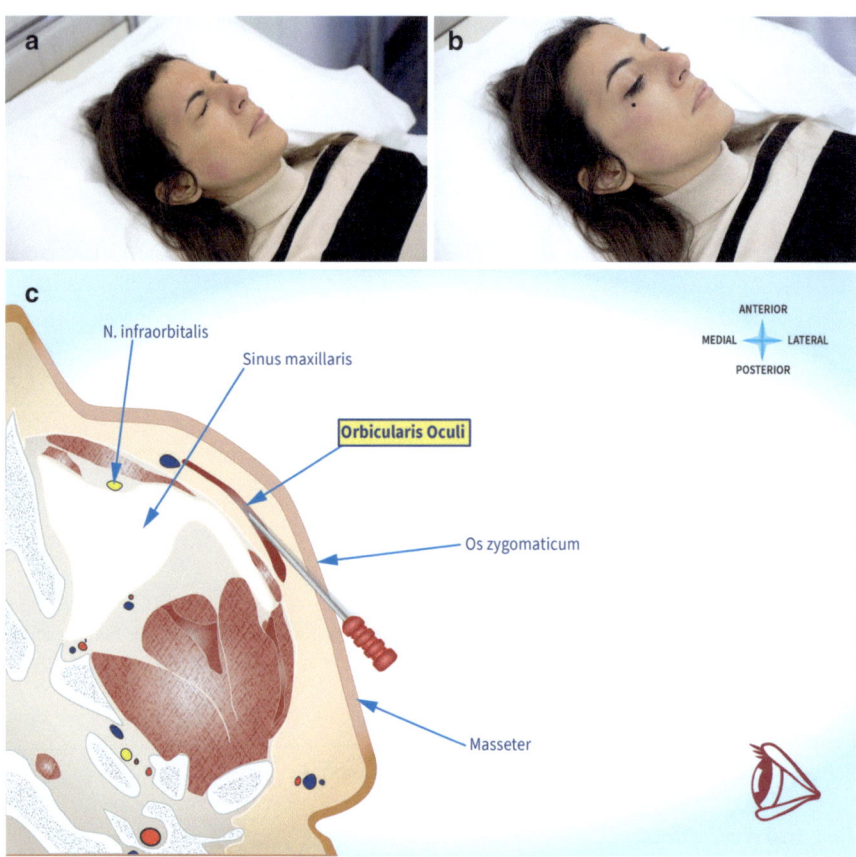

Fig. 8.4 Orbicularis oculi: (**a**) test, (**b**) insertion point, (**c**) cross-section below the orbit through the inferior margin of the temporal process of the zygomatic bone. The eye shows the perspective of the examiner

Innervation and Nerve Fibers Route
Facial nucleus (pons), facial nerve, temporal and zygomatic branch (Fig. 8.2).

Test and Activation
Ask the subject to close their eyelids (Fig. 8.4a).

Needle Insertion

With the subject lying supine and the head in a neutral position, insert the needle tangentially lateral to the inferior edge of the eye socket, with the needle pointing away from the eye (Fig. 8.4b).

Caveat: the muscle is very thin (Fig. 8.4c). If the electrode is inserted too perpendicular to the skin, it may enter the orbit and damage the eye. Anyway, soft tissue damage of the orbital area with swelling and/or hemorrhage may occur.

Comments

- The orbicularis oculi are tested in assessing Bell's palsy.
- In the orbicularis oculi, the duration of the motor unit action potential is normally briefer than in the limb muscles.
- The orbicularis oculi is the recording muscle for the blink reflex study.

8.2.3 Mentalis

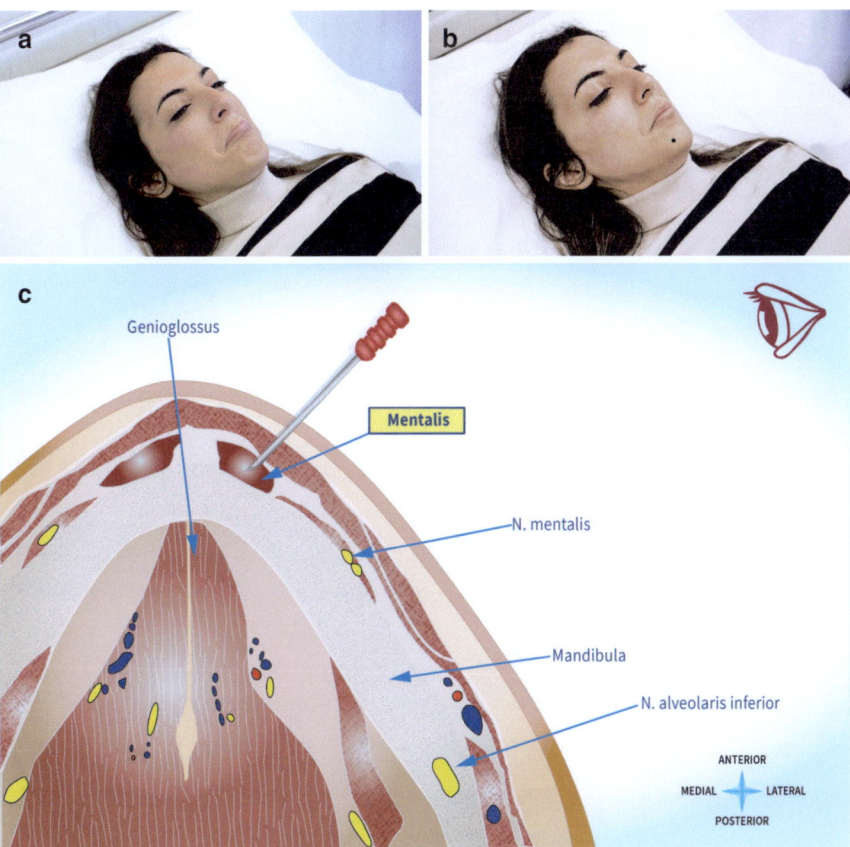

Fig. 8.5 Mentalis: (**a**) test, (**b**) insertion point, (**c**) cross-section through the body of the mandible just above the mental foramen. The eye shows the perspective of the examiner

Innervation and Nerve Fibers Route
Facial nucleus (pons), mandibular branch of facial nerve (Fig. 8.2).

Test and Activation
Ask the subject to raise the skin of the chin as in pouting (Fig. 8.5a).

Needle Insertion
With the subject lying supine and the head in a neutral position, insert the needle tangentially and superficially into the chin (Fig. 8.5b).
  Caveat: the muscle is very thin (Fig. 8.5c).

Comments

- The mentalis is tested in assessing Bell's palsy.
- The mentalis is useful to sample in patients with suspected motor neuron disease.
- In the mentalis, the duration of the motor unit action potential is normally briefer than in the limb muscles.

8.3 Spinal Accessory Nerve

8.3.1 Sternocleidomastoid

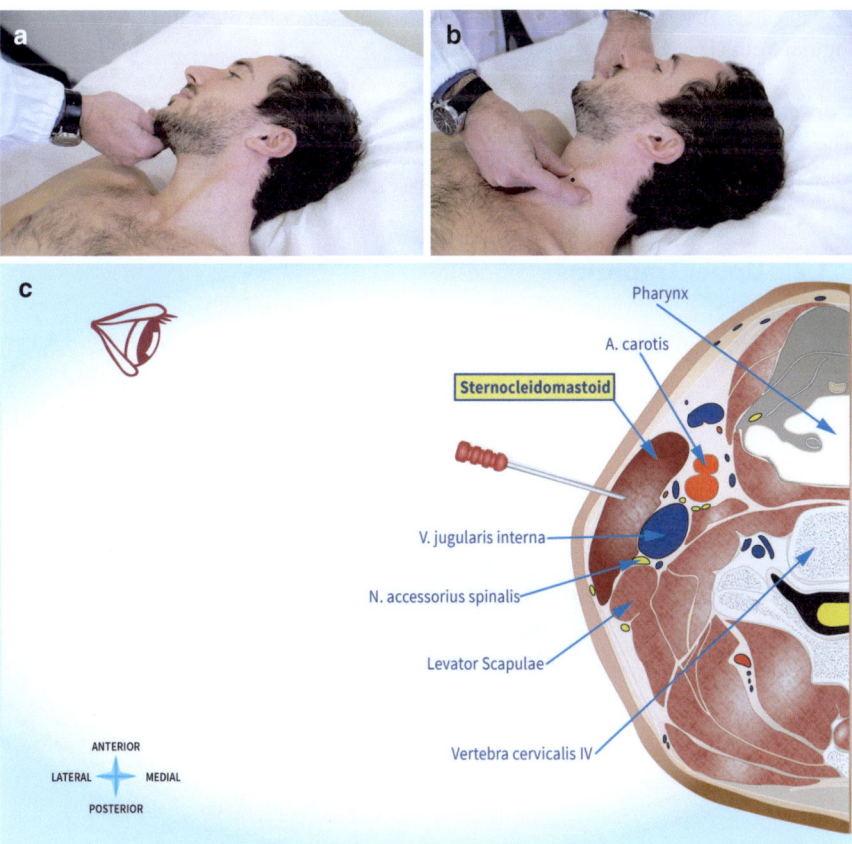

Fig. 8.6 Sternocleidomastoid: (**a**) test, (**b**) insertion point, (**c**) cross-section through the body, the transverse and spinous processes of the fourth cervical vertebra. The eye shows the perspective of the examiner

Innervation and Nerve Fibers Route
Spinal accessory nerve and C3 and C4 roots.

Test and Activation
Ask the subject to turn the head and neck to the opposite side of the muscle being tested. Apply pressure on the chin in the direction of contralateral rotation (Fig. 8.6a).

Needle Insertion

With the subject lying supine with the head in a neutral position, pinch and hold the muscle between the index finger and the thumb, and insert the needle at the midpoint of the muscle (at the level of the thyroid cartilage) (Fig. 8.6b).

Caveat: if inserted too deeply, the needle may puncture the carotid artery or the jugular vein (Fig. 8.6c).

Comment

The sternocleidomastoid is often involved in cervical dystonia (spasmodic torticollis).

8.3.2 Upper Trapezius

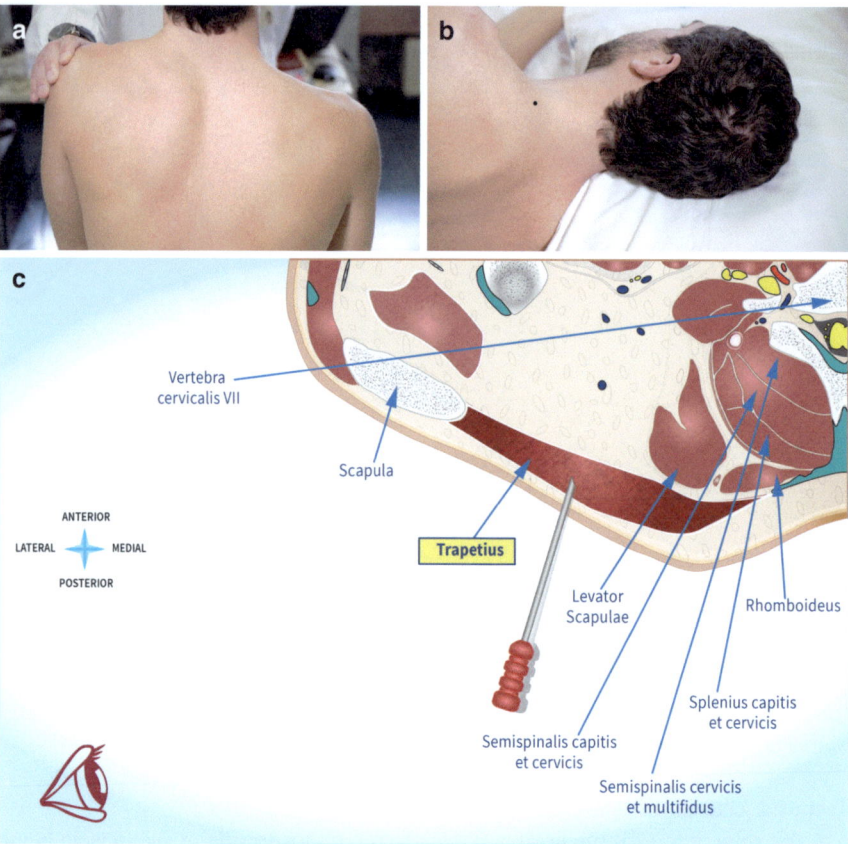

Fig. 8.7 Trapezius: (**a**) test, (**b**) insertion point, (**c**) cross-section through the upper margin of the body, posterior arch and transverse process of the seventh cervical vertebra. The eye shows the perspective of the examiner

Innervation and Nerve Fibers Route
Spinal accessory nerve and C3 and C4 roots.

Test and Activation
Ask the patient to shrug his/her shoulder. Apply pressure against the shoulder in the direction of depression (Fig. 8.7a).

Needle Insertion
With the subject lying on his/her side and the shoulder to be studied upward, insert the needle at the angle made by the neck and shoulder (Fig. 8.7b).

Caveat: the muscle is superficial (Fig. 8.7c). If the needle is inserted too medially and too deeply, it could reach the rhomboids or levator scapulae, both innervated by the dorsal scapular nerve, or the paraspinal muscles.

Comments
- The upper portion of the trapezius is the easiest to examine.
- The weakness of the trapezius induces abduction and medial rotation of the scapula (winging), which is more evident when abducting the arm.
- The trapezius is abnormal in spinal accessory nerve lesions caused by neck surgery and cervical lymph node biopsy, with sparing of the sternocleidomastoid muscle.

8.4 Hypoglossal Nerve, Musculi Linguae

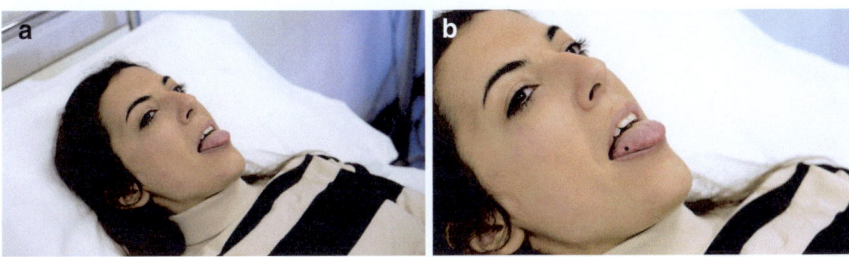

Fig. 8.8 Musculi linguae: (**a**) test, (**b**) insertion point

Innervation and Nerve Fibers Route
Nucleus of the 12th cranial nerve (medulla), hypoglossal nerve.

Test and Activation
Ask the subject to stick their tongue out of their mouth (Fig. 8.8a).

Needle Insertion
Ask the subject to protrude their tongue, hold the end of their tongue with a gauze pad, and then insert the needle laterally into the side of the tongue (Fig. 8.8b).

Comments

- When the tongue muscles of both sides have unequal strength, the tip deviates toward the weak side when protruded (it points to the side of the lesion).
- The tongue is difficult to relax. To assess spontaneous activity at EMG, ask the subject, after the needle insertion, to pull their tongue back on the floor of the mouth and try to relax.
- In the tongue, the duration of the motor unit action potential is normally briefer than in the limb muscles.
- The tongue is useful to sample in patients with suspected motor neuron disease.

Muscles Innervated by Nerves of the Brachial Plexus

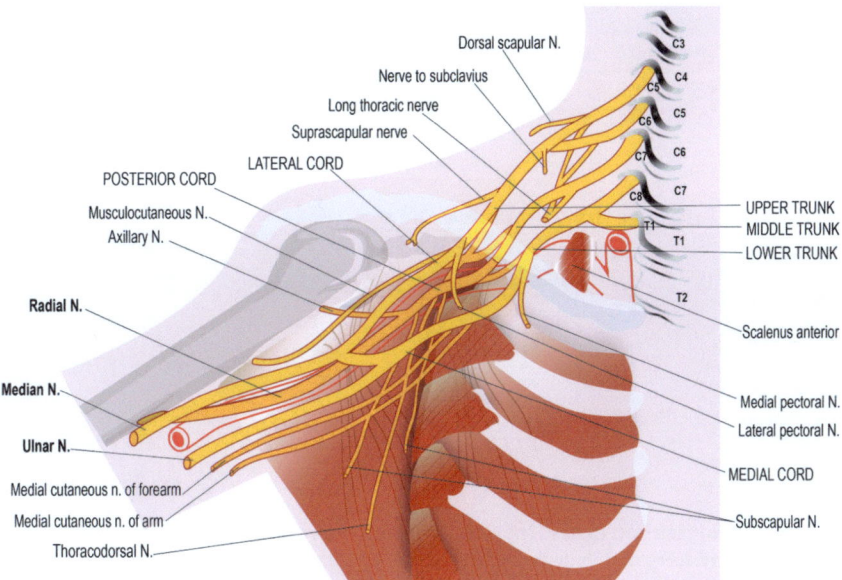

Fig. 9.1 The brachial plexus and its branches

9.1 Long Thoracic Nerve, Serratus Anterior

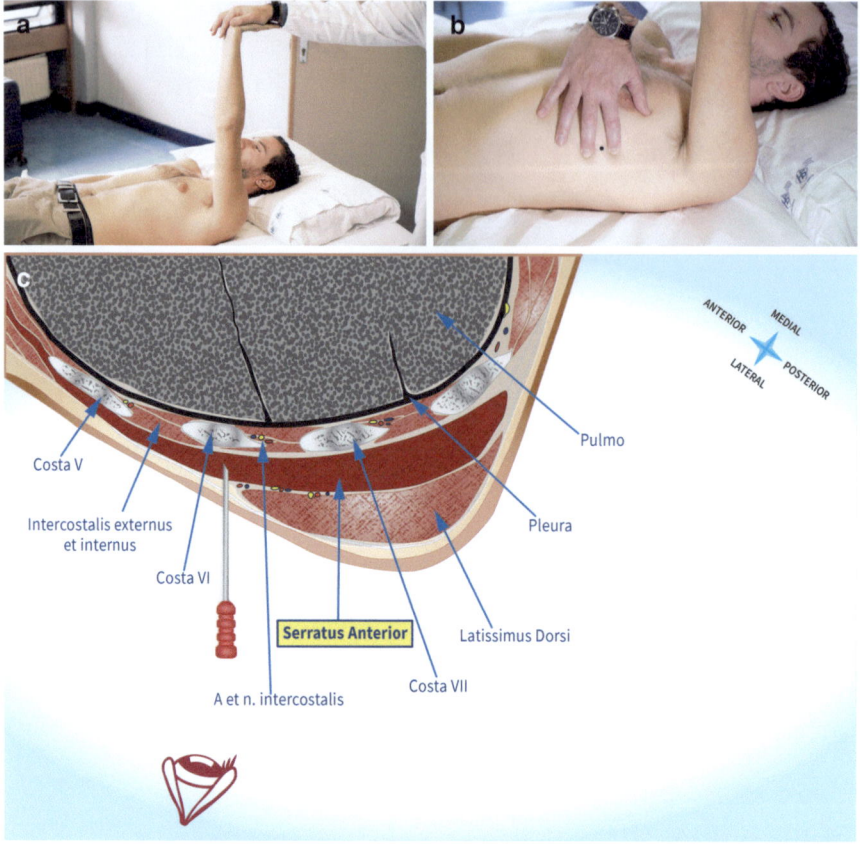

Fig. 9.2 Serratus anterior: (**a**) test, (**b**) insertion point, (**c**) cross-section through the middle of the eight thoracic vertebra and through the fifth, sixth, seventh and eighth ribs. The eye shows the perspective of the examiner

Innervation and Nerve Fibers Route
C5, C6, and C7 roots, long thoracic nerve (Fig. 9.1).

Test and Activation
With the subject lying supine and the arm straight upward, ask to push while applying pressure against the palm of the hand and pressing downward (Fig. 9.2a).

Needle Insertion
With the subject lying supine, insert the needle over the sixth rib at the mid-axillary line while covering with two fingers of the other hand the nearby intercostal spaces (Fig. 9.2b).

Caveat: if the needle is inserted between the ribs, there is a risk of causing pneumothorax or injuring the neurovascular bundle (Fig. 9.2c).

Comments

- The serratus anterior helps to fix and stabilize the scapula against the chest wall.
- When the serratus anterior is paralyzed, the patient has a winged scapula with the vertebral border that stands out, especially when trying to push (for instance, against a wall).
- The serratus anterior is difficult to study because most of the muscle lies between the rib cage and the scapula.
- The long thoracic nerve arises directly from roots C5, C6, and C7, and the serratus anterior is spared in lesions of the brachial plexus.
- The serratus anterior is often involved, even solely, in neuralgic amyotrophy (acute brachial plexitis).

9.2 Dorsal Scapular Nerve, Rhomboideus Major

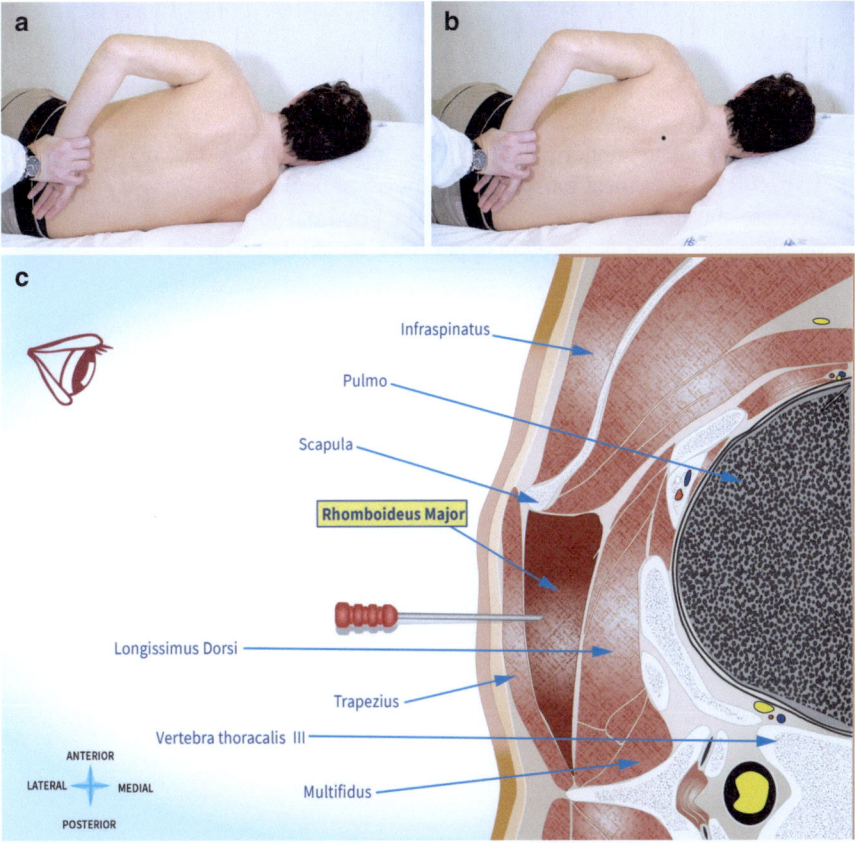

Fig. 9.3 Rhomboideus major: (**a**) test, (**b**) insertion point, (**c**) cross-section passing through the upper margin of the body, the arch and the spinous process of the third thoracic vertebra. The eye shows the perspective of the examiner

Innervation and Nerve Fibers Route
C4, C5, dorsal scapular nerve (Fig. 9.1).

Test and Activation
With the subject lying on his/her side, the arm internally rotated, and the dorsum of the hand resting on the small back, ask to raise the hand from the small back (Fig. 9.3a).

Needle Insertion
With the subject lying on his/her side and the side to be studied upward, insert the needle midway between the spine of the scapula and the inferior angle of the scapula just medial to the vertebral border (Fig. 9.3b).

Caveat: if the needle is inserted too superficially, it will be in the trapezius innervated by the accessory nerve; if inserted too deeply, it will be in the thoracic paraspinal muscles (Fig. 9.3c). Pneumothorax has been reported rarely with too deep needle insertion.

Comments

- When the rhomboideus major is weak, the scapula is winged with the inferior angle rotated outward.
- The rhomboideus major is prevalently innervated from the C5 nerve root. Therefore, it is an ideal muscle to test for C5 radiculopathy.
- Because the dorsal scapular nerve arises proximal to the brachial plexus, the rhomboideus major is spared in upper trunk brachial plexopathy.

9.3 Medial and Lateral Pectoral Nerves, Pectoralis Major

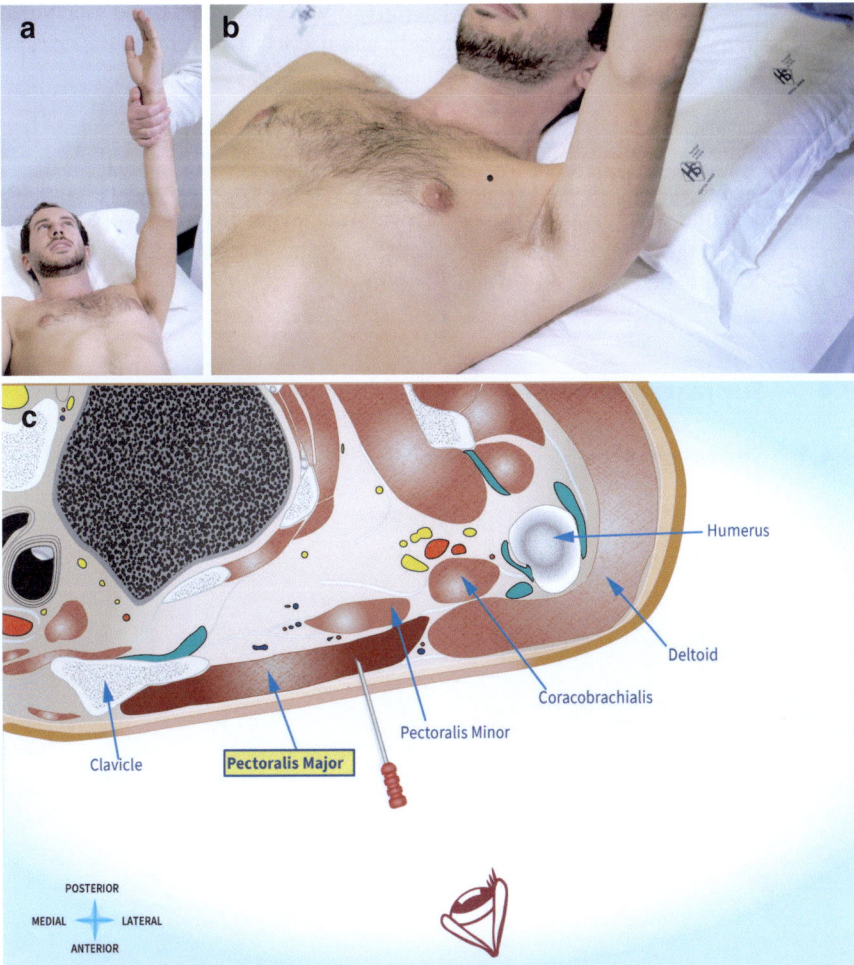

Fig. 9.4 Pectoralis major: (**a**) test, (**b**) insertion point, (**c**) cross-section passing through the upper margin of the body of the third thoracic vertebra, the sternal end of the clavicle, and the surgical neck of the humerus. The eye shows the perspective of the examiner

Innervation and Nerve Fibers Route
Clavicular portion: C5, C6, upper trunk, lateral cord, lateral pectoral nerve. Sternocostal portion: C7, C8, T1, middle and lower trunk, medial cord, medial pectoral nerve (Fig. 9.1).

Test and Activation
With the subject lying supine, the elbow extended, and the shoulder at 90°, ask to adduct the arm. Apply pressure against the forearm in the direction of abduction (Fig. 9.4a).

Needle Insertion

With the subject lying supine, insert the needle into the anterior axillary fold (Fig. 9.4b).

Caveat: if the needle is inserted too laterally, it will be in the deltoid innervated by the axillary nerve; if inserted too laterally and deeply, it may reach the coracobrachialis innervated by the musculocutaneous nerve (Fig. 9.4c).

Comments

- The pectoralis major is innervated by all roots and trunks of the brachial plexus.
- The clavicular portion (superior) of the pectoralis major may be involved in lateral cord and C5 and C6 lesions; the sternocostal portion (inferior) may be involved in medial cord and C8 and T1 lesions.

9.4 Suprascapular Nerve

9.4.1 Supraspinatus

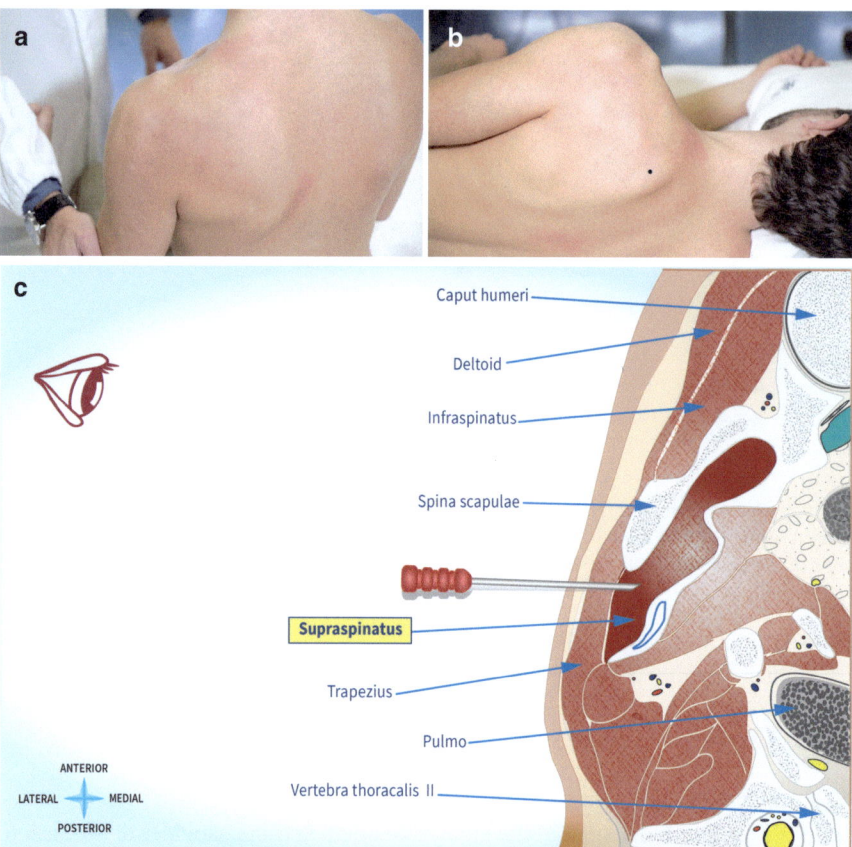

Fig. 9.5 Supraspinatus: (**a**) test, (**b**) insertion point, (**c**) cross-section through the transverse and spinous process of the second thoracic vertebra and supraspinous fossa of the scapula. The eye shows the perspective of the examiner

9.4 Suprascapular Nerve

Innervation and Nerve Fibers Route

C5, C6, upper trunk, suprascapular nerve (Fig. 9.1).

Test and Activation

Ask the patient to abduct the shoulder with the elbow flexed. Apply pressure against the distal humerus. Although the supraspinatus is responsible for the first few degrees of arm abduction, both the supraspinatus and deltoid act concurrently in completing the shoulder abduction, and their individual actions cannot be distinguished (Fig. 9.5a).

Needle Insertion

With the subject lying on his/her side and the shoulder to be studied upward, insert the needle into the supraspinous fossa just above the middle of the spine of the scapula (Fig. 9.5b). The electrode will pass through the mid-trapezius muscle.

Caveat: if the needle is inserted too superficially, it will be in the trapezius innervated by the accessory nerve (Fig. 9.5c). Pneumothorax has been rarely reported with needle insertion too rostral and deep.

Comments

- The supraspinatus may be involved in suprascapular nerve entrapment at the scapular notch, in upper trunk plexopathy, and in C5 and C6 radiculopathy.
- The supraspinatus is spared in suprascapular neuropathy at the spinoglenoid notch.

9.4.2 Infraspinatus

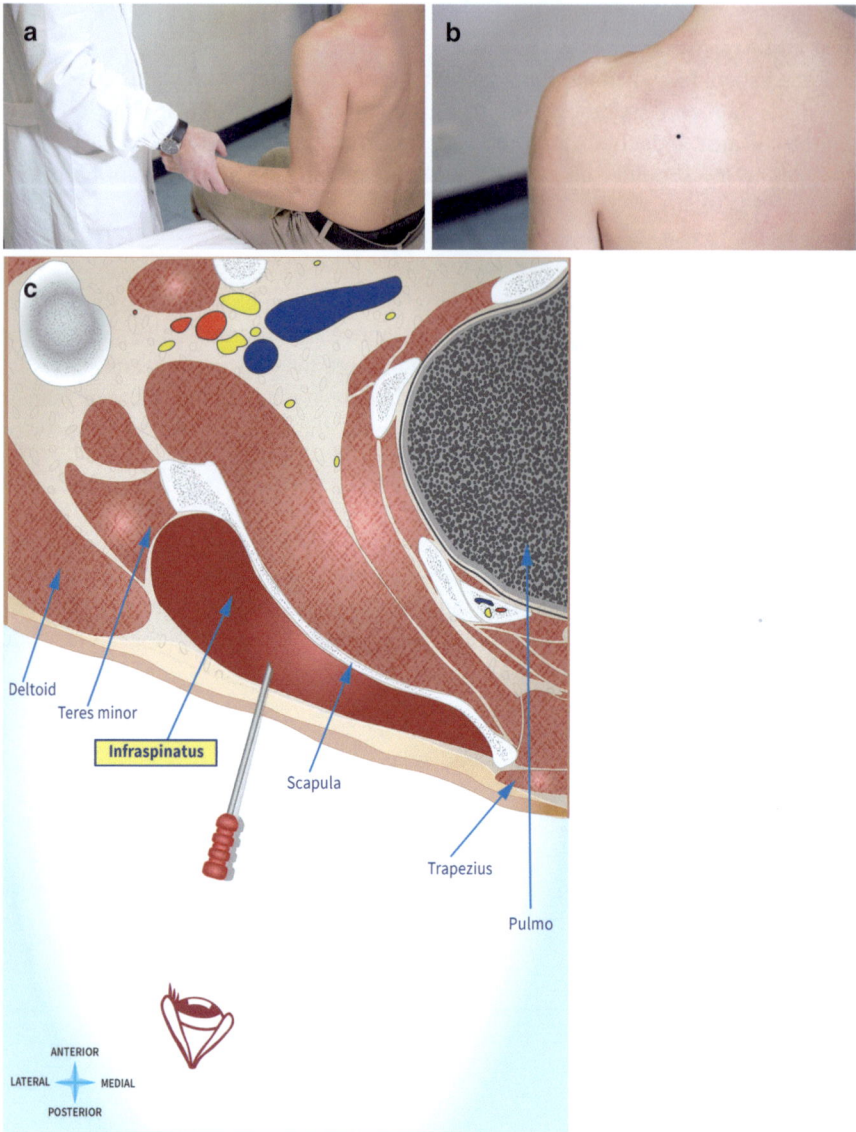

Fig. 9.6 Infraspinatus: (**a**) test, (**b**) insertion point, (**c**) cross-section trough the scapula about 1 cm below the lower margin of the glenoid fossa. The eye shows the perspective of the examiner

Innervation and Nerve Fibers Route
C5, C6, upper trunk, suprascapular nerve (Fig. 9.1).

9.4 Suprascapular Nerve

Test and Activation

With the subject seated and the elbow flexed at 90°, ask to externally rotate the arm (Fig. 9.6a).

Needle Insertion

With the subject seated, insert the needle one to two fingerbreadths below the midpoint of the scapular spine until it reaches the bone, and then withdraw it slightly (Fig. 9.6b).

Caveat: if the needle is inserted too upward and laterally, it will be in the posterior deltoid innervated by the accessory nerve (Fig. 9.6c).

Comments

- In testing the infraspinatus, there is no risk of pneumothorax as the muscle lies over the scapular bone.
- The infraspinatus, as well as the supraspinatus, may be involved in suprascapular nerve entrapment at the suprascapular notch.
- The infraspinatus is exclusively involved in suprascapular nerve entrapment at the spinoglenoid notch. It may be involved also in upper trunk plexopathy and C5 and C6 radiculopathy.

9.5 Thoracodorsal Nerve, Latissimus Dorsi

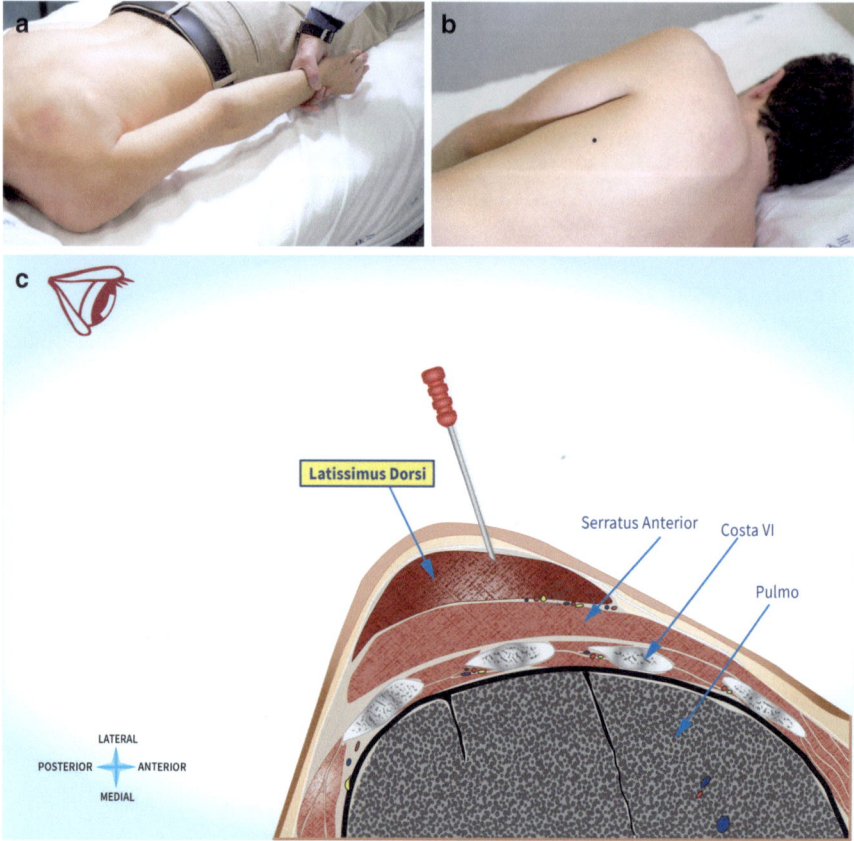

Fig. 9.7 Latissimus dorsi: (**a**) test, (**b**) insertion point, (**c**) cross-section through the middle of the eight thoracic vertebra and through the fifth, sixth, seventh and eighth ribs. The eye shows the perspective of the examiner

Innervation and Nerve Fibers Route
C6, C7, and C8; upper, middle, and lower trunks; posterior cord; thoracodorsal nerve (Fig. 9.1).

Test and Activation
With the subject prone and the arm at the side, ask to internally rotate, adduct, and extend the arm (Fig. 9.7a).

Needle Insertion
With the patient lying on his/her side and the side to be studied placed upward, insert the needle lateral to the tip of the scapula at the posterior axillary line (Fig. 9.7b).

9.5 Thoracodorsal Nerve, Latissimus Dorsi

Caveat: if the needle is inserted too deeply, it will be in the serratus anterior innervated by the long thoracic nerve; if it is inserted too medially, it will be in the teres major innervated by the subscapular nerve (Fig. 9.7c).

Comments

- The latissimus dorsi is more difficult to study than the triceps and other radial-innervated muscles that have similar root, trunk, and cord innervation.
- However, testing the latissimus dorsi may help to localize the lesion to the posterior cord proximally to the axillary nerve.

10

Muscles Innervated by the Musculocutaneous, Axillary, and Radial Nerves

10.1 Musculocutaneous Nerve, Biceps Brachii

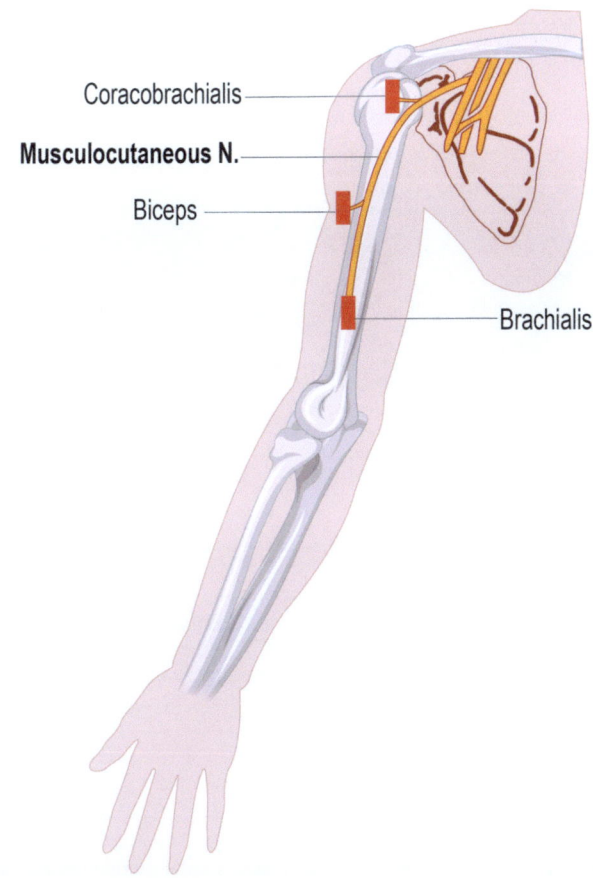

Fig. 10.1 The musculocutaneous nerve and the muscles it supplies

© Società Italiana di Neurofisiologia Clinica 2025
A. Uncini et al., *A Companion to Peripheral Nervous System Examination*,
https://doi.org/10.1007/978-3-031-63628-8_10

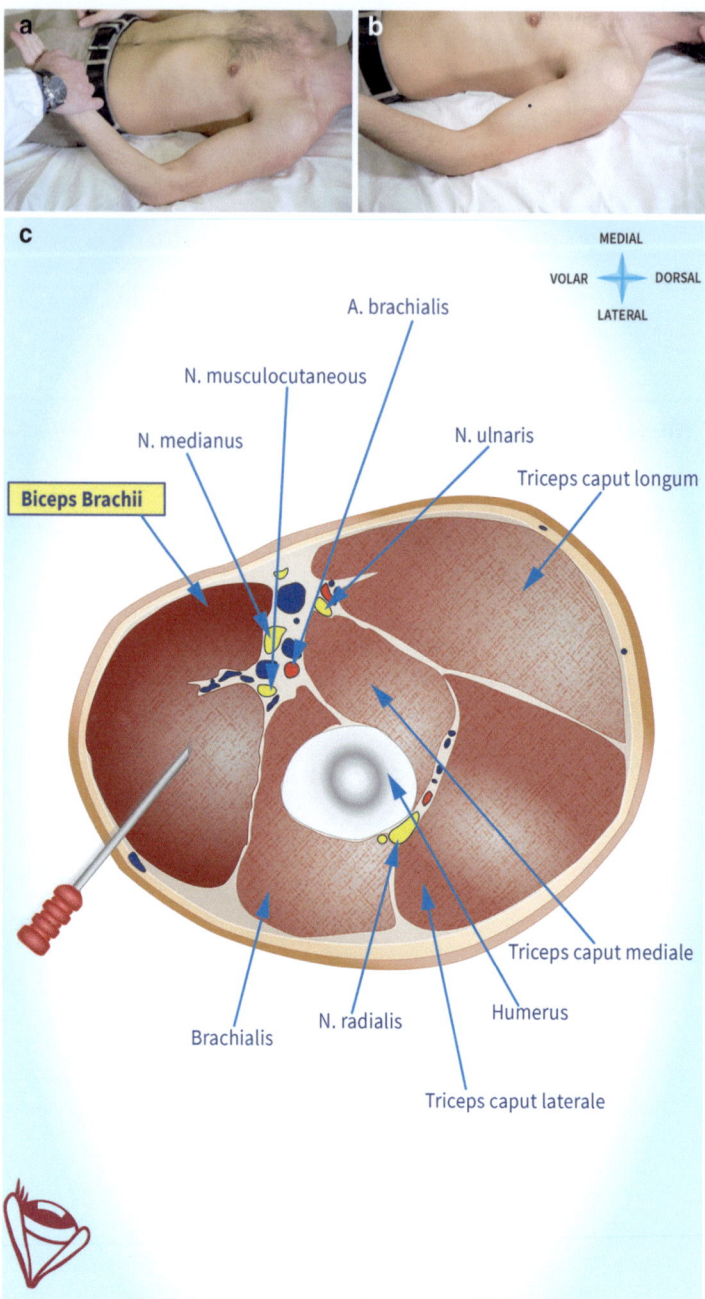

Fig. 10.2 Biceps brachii: (**a**) test, (**b**) insertion point, (**c**) cross-section through about the midshaft of the humerus. The eye shows the perspective of the examiner

10.1 Musculocutaneous Nerve, Biceps Brachii

Innervation and Nerve Fibers Route
C5, C6, upper trunk, lateral cord, musculocutaneous nerve (Fig. 10.1).

Test and Activation
With the subject's forearm supinated, ask to flex the elbow. Apply pressure against the lower forearm (Fig. 10.2a).

Needle Insertion
With a lateral approach, insert the needle at the midpoint between the biceps tendon and the shoulder (Fig. 10.2b).

Caveat: if sampled from the medial side, the brachial artery, the median nerve, and large veins might be at risk of injury (Fig. 10.2c).

Comments

- The biceps brachii has a dual function: a powerful elbow flexor and a strong supinator of the forearm.
- The biceps brachii is the most accessible muscle innervated by the musculocutaneous nerve.
- The biceps brachii is used as the recording muscle for the musculocutaneous nerve motor conduction study.
- The biceps brachii may be affected in lesions of the musculocutaneous nerve, upper trunk, lateral cord, and C5 and C6.

10.2 Axillary Nerve

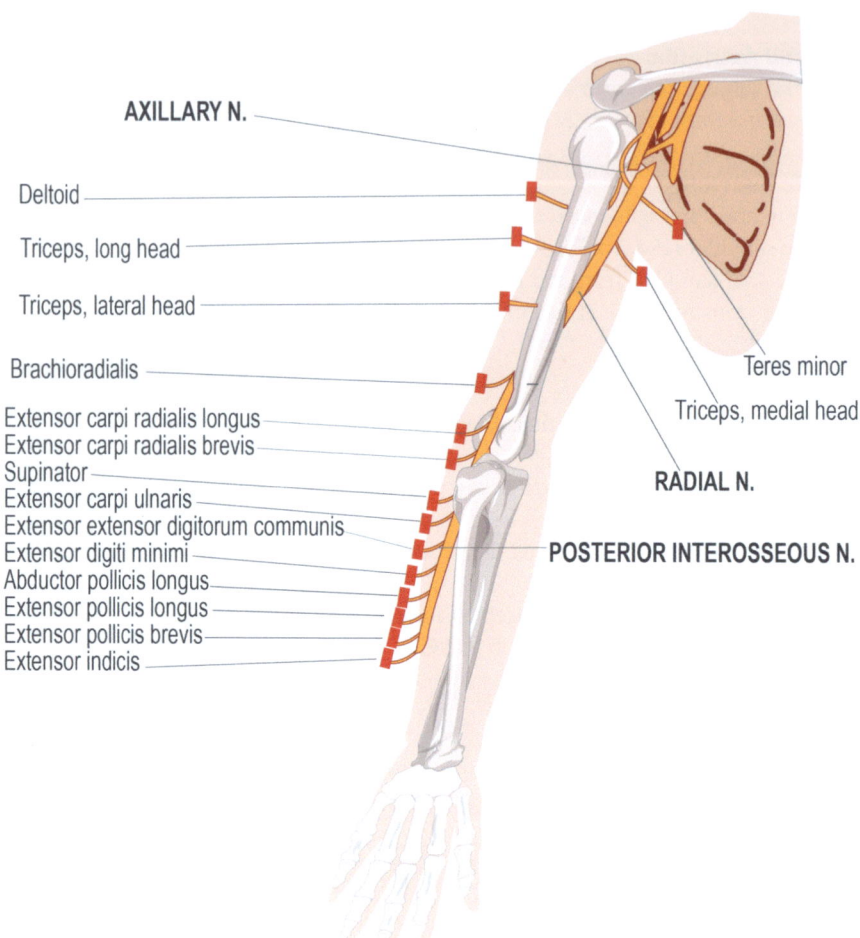

Fig. 10.3 The axillary and radial nerves and the muscles they supply

10.2 Axillary Nerve

10.2.1 Teres Minor

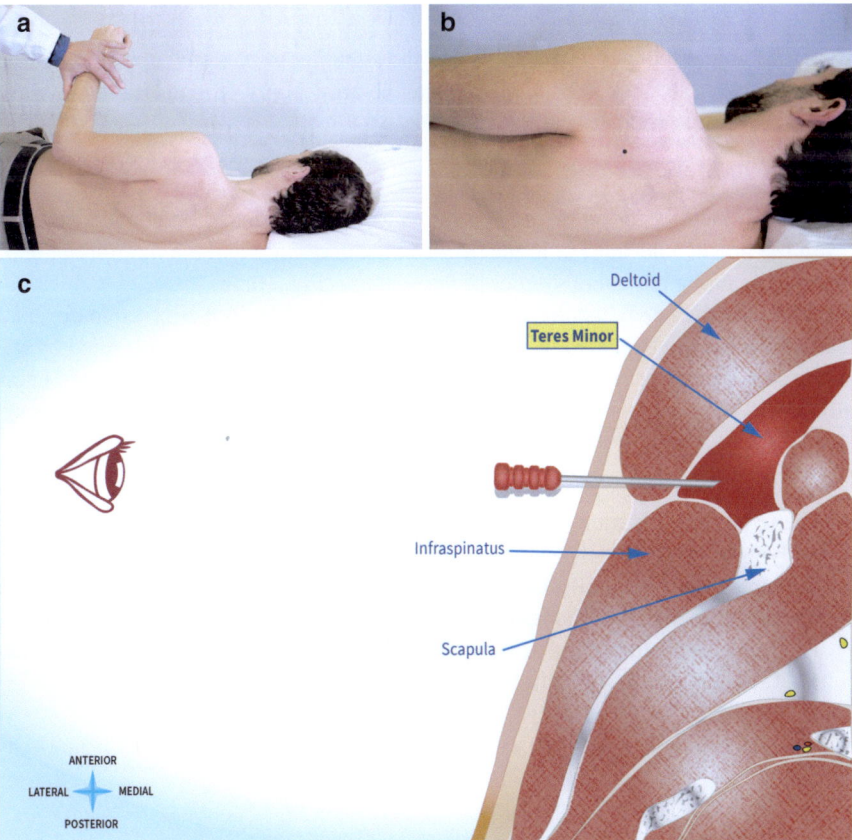

Fig. 10.4 Teres minor: (**a**) test, (**b**) insertion point, (**c**) cross-section through the upper margin of the body of the third thoracic vertebra and through the scapula about 1 cm below the lower margin of the glenoid fossa. The eye shows the perspective of the examiner

Innervation and Nerve Fibers Route
C5, C6, upper trunk, posterior cord, axillary nerve (Fig. 10.3).

Test and Activation
With the subject lying on his/her side and the shoulder to be tested upward with the elbow flexed at 90°, ask to externally rotate the arm. Apply pressure on the forearm in the direction to rotate the humerus medially (Fig. 10.4a).

Needle Insertion
With the subject lying on his or her side and the shoulder to be studied upward, insert the needle one-third of a line between the acromion and the inferior angle of the scapula along the lateral border of the scapula (Fig. 10.4b).

Caveat: if the needle is inserted too medially, it will be in the infraspinatus, and if it is inserted too cephalad, it will be in the supraspinatus, both innervated by the suprascapular nerve (Fig. 10.4c). If the needle is inserted too superficially and laterally, it will be in the posterior head of the deltoid, innervated by the axillary nerve.

Comments

- The teres minor is more difficult to test than the deltoid, which is preferred for studying an axillary neuropathy.
- The teres minor is usually spared in lesions of the axillary nerve secondary to the fracture of the neck of the humerus or dislocation of the glenohumeral joint.
- The teres minor may be also involved in upper trunk/posterior cord plexopathy and C5 and C6 radiculopathy.

10.2.2 Deltoid

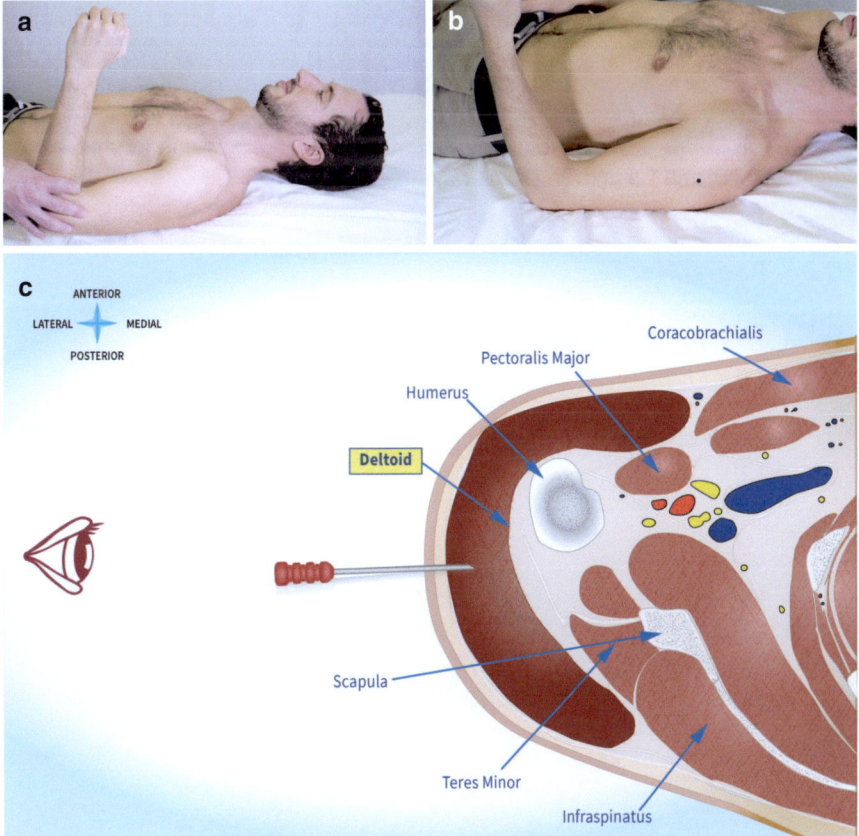

Fig. 10.5 Deltoid: (**a**) test, (**b**) insertion point, (**c**) cross-section through the upper margin of the body of the third thoracic vertebra, through the scapula about 1 cm below the lower margin of the glenoid fossa and the surgical neck of the humerus. The eye shows the perspective of the examiner

Innervation and Nerve Fibers Route
C5, C6, upper trunk, posterior cord, axillary nerve (Fig. 10.3).

Test and Activation
Ask the subject to abduct the shoulder with the elbow flexed. Apply pressure against the distal end of the humerus (Fig. 10.5a).

Needle Insertion
Halfway between the acromion and the deltoid insertion on the humerus (Fig. 10.5b, c).

Comments

- The medial head of the deltoid is the easiest of the three deltoid heads to study.
- The deltoid is used as the recording muscle in the axillary nerve motor conduction study.
- In normal subjects, the percentage of polyphasic motor unit action potentials is higher (up to 20%) in the deltoid than in other muscles.
- The deltoid may be involved in axillary neuropathy, upper trunk/posterior cord plexopathy, and C5 and C6 radiculopathy.

10.3 Radial Nerve

10.3.1 Triceps Brachii-Caput Laterale

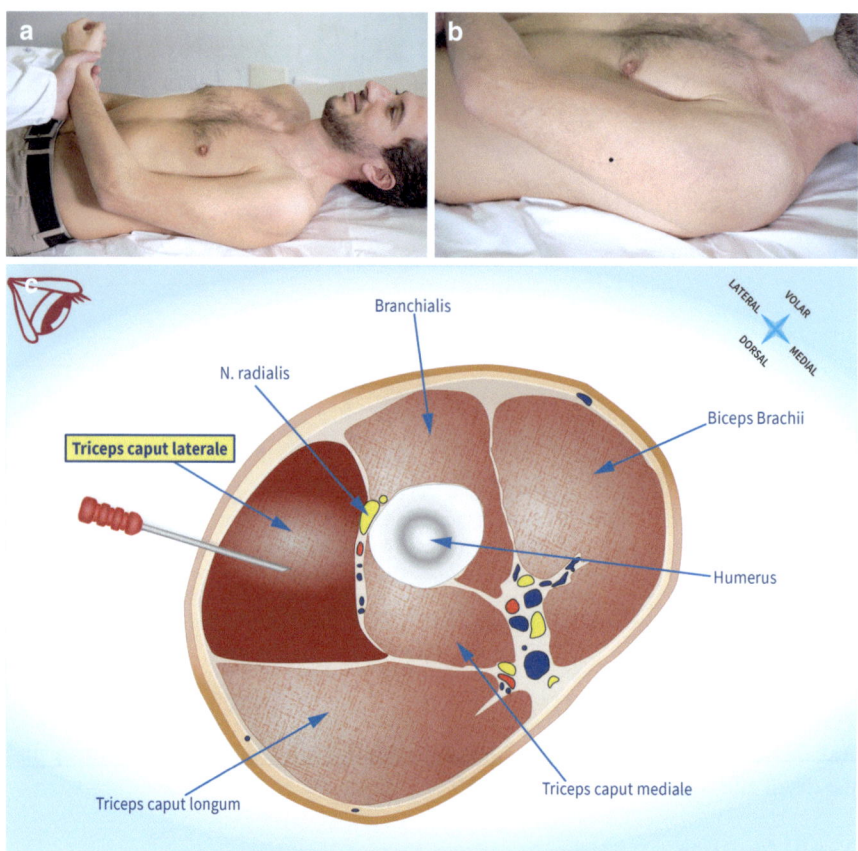

Fig. 10.6 Triceps-caput laterale: (**a**) test, (**b**) insertion point, (**c**) cross-section through about the half of the humerus. The eye shows the perspective of the examiner

10.3 Radial Nerve

Innervation and Nerve Fibers Route
C6, C7, C8, T1, middle and lower trunks, posterior cord, radial nerve (Fig. 10.3).

Test and Activation
Extend the forearm at the elbow, and apply pressure against the lower forearm in the direction of flexion (Fig. 10.6a).

Needle Insertion
With the subject's forearm pronated and the elbow flexed, insert the needle just below the midpoint of the muscle between the lateral epicondyle and the shoulder (Fig. 10.6b).

Caveat: as long as the triceps brachii is sampled from the lateral approach, there are no other vascular structures or major nerves that can be injured (Fig. 10.6c).

Comments

- The lateral head of the triceps is the easiest to study of the three heads.
- The triceps, because of its proximal innervation by the radial nerve, is spared in radial neuropathy at the spiral groove ("crutch paralysis" and "Saturday night palsy").
- The triceps may be involved in C7 radiculopathy.

10.3.2 Brachioradialis

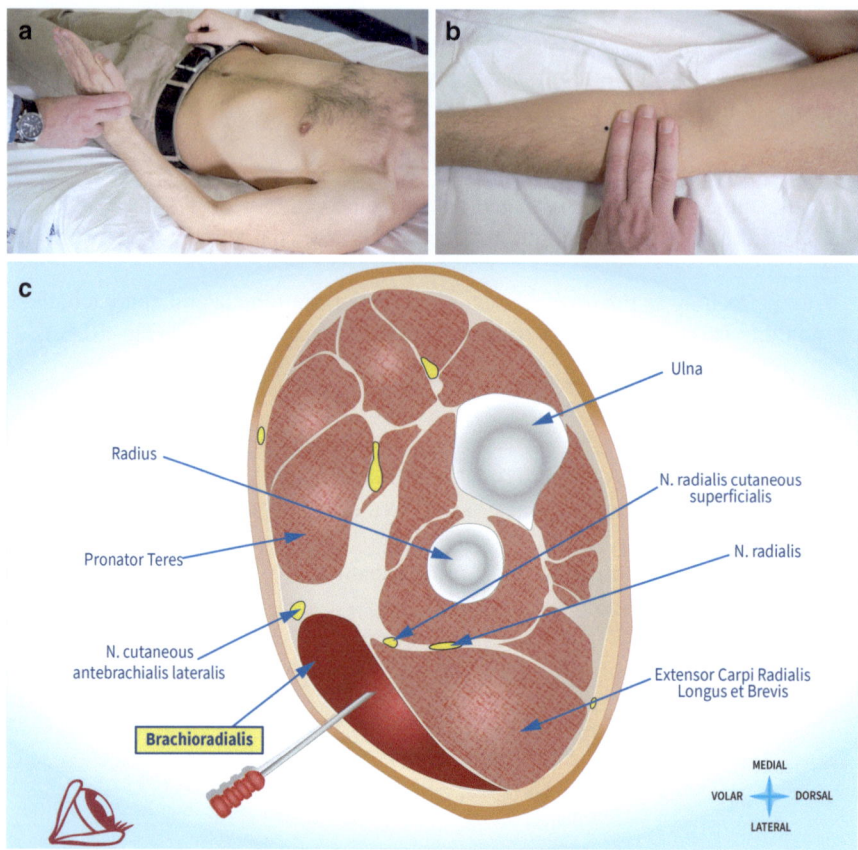

Fig. 10.7 Brachioradialis: (**a**) test, (**b**) insertion point, (**c**) cross-section through the tuberosity of the ulna and the neck of the radius. The eye shows the perspective of the examiner

Innervation and Nerve Fibers Route
C5, C6, C7, upper trunk, posterior cord, radial nerve (Fig. 10.3).

Test and Activation
With the subject's forearm in a neutral position, ask to flex the elbow. Apply pressure against the lower forearm in the direction of extension (Fig. 10.7a).

Needle Insertion
Three fingerbreadths distal to the midpoint between the biceps tendon and the lateral epicondyle (Fig. 10.7b).

Caveat: if the needle is inserted too laterally, it will be in the extensor carpi radialis (ECR), which is also innervated by the radial nerve (Fig. 10.7c).

Comments

- The brachioradialis may be affected in upper trunk plexopathy or C5 and C6 radiculopathy.
- The brachioradialis is affected in radial nerve injuries at the spiral groove or above.
- The brachioradialis is spared in the posterior interosseous syndrome.

10.3.3 Extensor Carpi Radialis (ECR) Longus

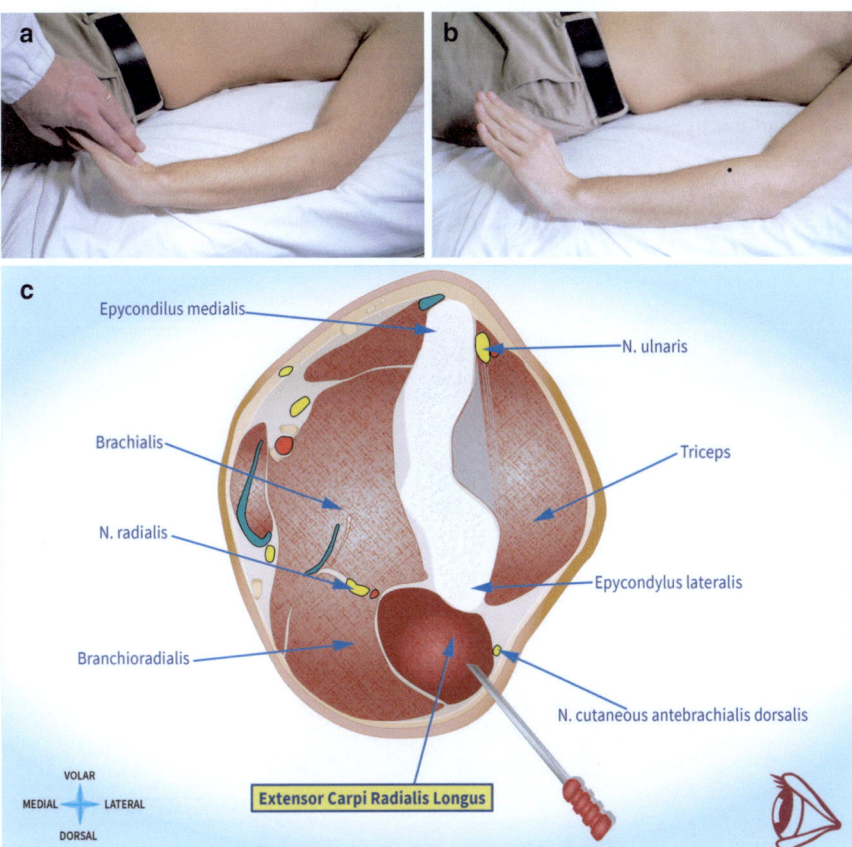

Fig. 10.8 Extensor carpi radialis longus: (**a**) test, (**b**) insertion point, (**c**) cross-section through the medial and lateral epicondyle of the humerus. The eye shows the perspective of the examiner

Innervation and Nerve Fibers Route
C6, C7, upper and middle trunks, posterior cord, radial nerve (Fig. 10.3).

Test and Activation
With the subject's forearm pronated, ask to extend the wrist toward the radial side. Apply pressure against the dorsum of the hand along the second metacarpal bone in the direction of flexion toward the ulnar side (Fig. 10.8a).

Needle Insertion

Two fingerbreadths distal to the lateral epicondyle (Fig. 10.8b). Because of anatomical position and identical function, it is difficult to examine separately the ECR longus and brevis.

Caveat: if the needle is inserted too medially, it will be in the brachioradialis; if inserted too distally, it may be in other wrist or finger extensor muscles, all innervated by the radial nerve (Fig. 10.8c).

Comments

- The ECR is spared in posterior interosseous syndrome.
- The ECR may be involved in lesions of the radial nerve at or above the spiral groove.
- The ECR may be involved in upper trunk plexopathy and C6 and C7 radiculopathy.

10.3.4 Extensor Digitorum Communis (EDC)

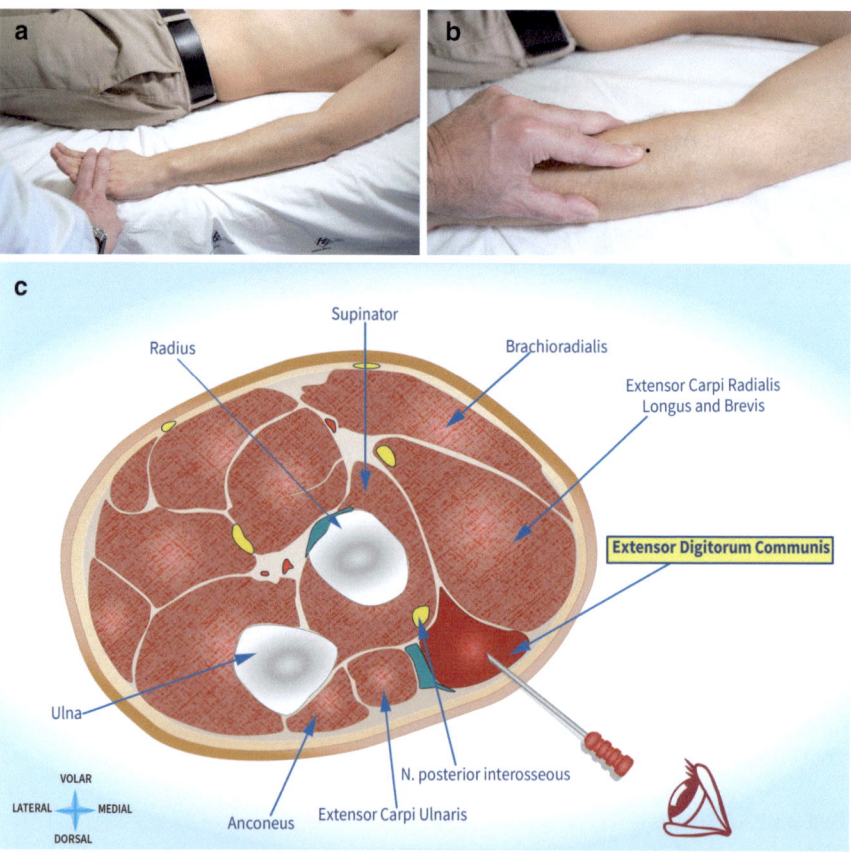

Fig. 10.9 Extensor digitorum communis: (**a**) test, (**b**) insertion point, (**c**) cross-section through the upper third of the forearm. The eye shows the perspective of the examiner

10.3 Radial Nerve

Innervation and Nerve Fibers Route

C7, C8, middle and lower trunks, posterior cord, radial nerve, posterior interosseus nerve (Fig. 10.3).

Test and Activation

Stabilizing the subject's wrist, ask to extend the proximal phalanges of the second through fifth digits. Apply pressure against the dorsal surfaces of the proximal phalanges in the direction of flexion (Fig. 10.9a).

Needle Insertion

Hold the forearm pronated at the junction of the upper and middle third with the thumb and middle finger on the radius and ulna, and then insert the needle at the tip of the index finger (Fig. 10.9b).

Caveat: if the needle is inserted too deeply, it will be in the supinator or extensor pollicis longus; if inserted too medially, it will be in the extensor carpi radialis; if inserted too laterally, it will be in the extensor carpi ulnaris (Fig. 10.9c). All these muscles are innervated by the radial nerve. If the needle is inserted too deeply, it may reach the posterior interosseus nerve.

Comments

- The extensor digitorum communis (EDC) is superficial and easily palpated when the patient activates the muscle.
- The EDC may be affected in all radial nerve lesions, including posterior interosseous syndrome.
- The EDC is used for single-fiber electromyography.

10.3.5 Extensor Indicis Proprius (EIP)

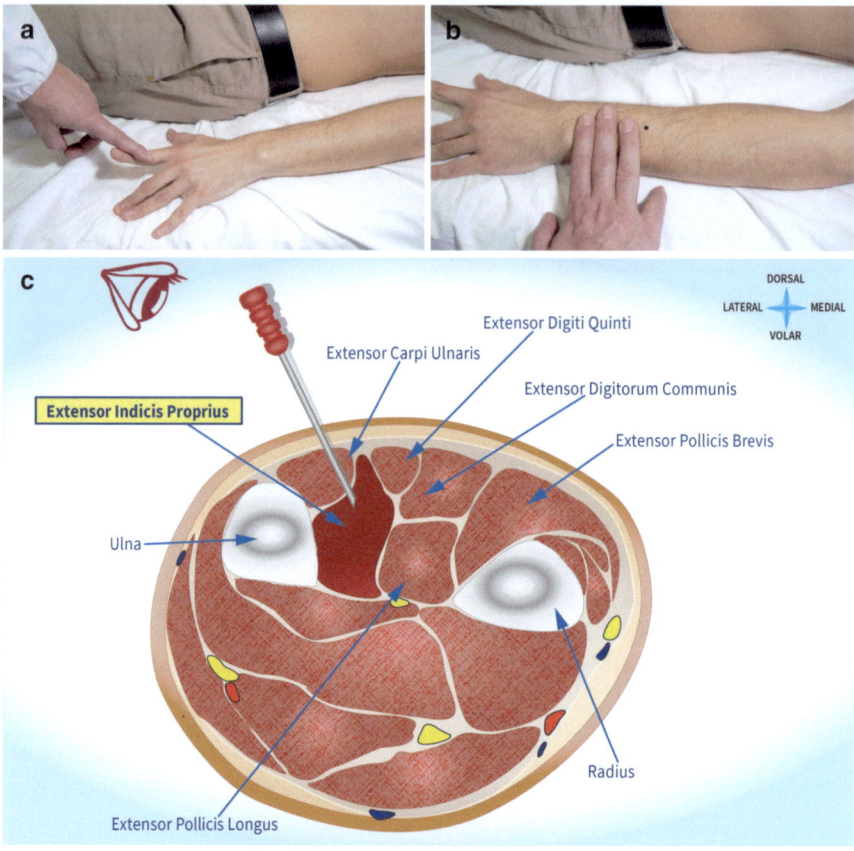

Fig. 10.10 Extensor indicis proprius: (**a**) test, (**b**) insertion point, (**c**) cross-section through the middle of the forearm. The eye shows the perspective of the examiner

10.3 Radial Nerve

Innervation and Nerve Fibers Route

C7, C8, middle and lower trunks, posterior cord, radial nerve, posterior interosseus nerve (Fig. 10.3).

Test and Activation

With the subject's forearm fully pronated, ask to extend the index finger. Apply pressure to the first phalanx (Fig. 10.10a).

Needle Insertion

With the forearm pronated, insert the needle three fingerbreadths proximal to the ulnar styloid just radial to the ulna (Fig. 10.10b).

Caveat: if the needle is inserted too superficially, it will be in the extensor carpi ulnaris or extensor digiti minimi, both muscles innervated by the radial nerve (Fig. 10.10c).

Comments

- The extensor indicis proprius (EIP) is the most distal radial-innervated muscle.
- The EIP is used as the recording muscle in radial nerve motor conduction studies.
- The EIP may be involved in all radial nerve lesions, including posterior interosseous nerve palsy.
- The EIP may be involved in lower trunk/posterior cord plexopathy and C8 radiculopathy.

Muscles Innervated by the Median Nerve 11

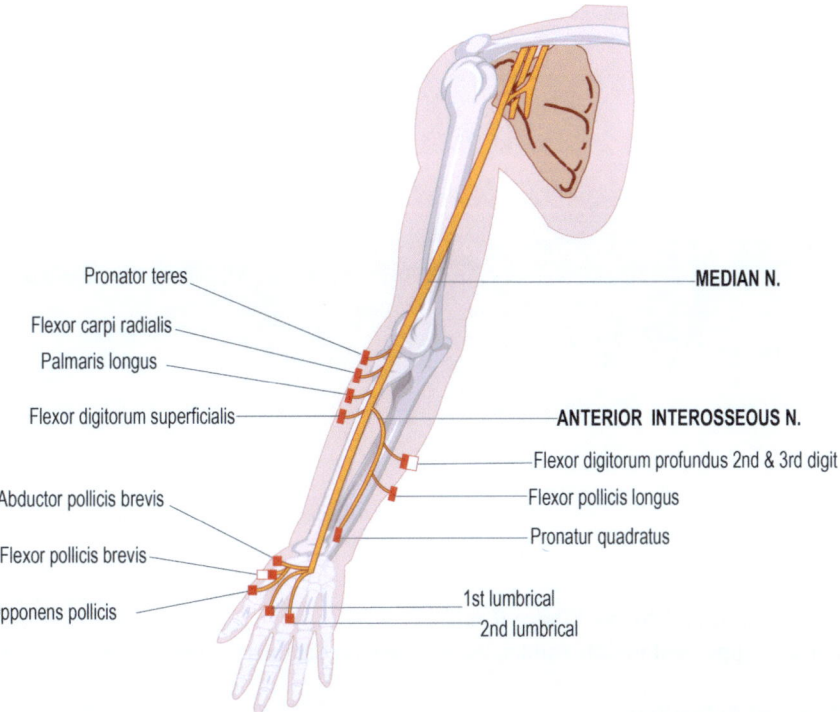

Fig. 11.1 Median nerve and the muscles it supplies. The white part of the label of the flexor digitorum profundus and flexor pollicis brevis indicates that these muscles are innervated also from the ulnar nerve

© Società Italiana di Neurofisiologia Clinica 2025
A. Uncini et al., *A Companion to Peripheral Nervous System Examination*,
https://doi.org/10.1007/978-3-031-63628-8_11

11.1 Pronator Teres

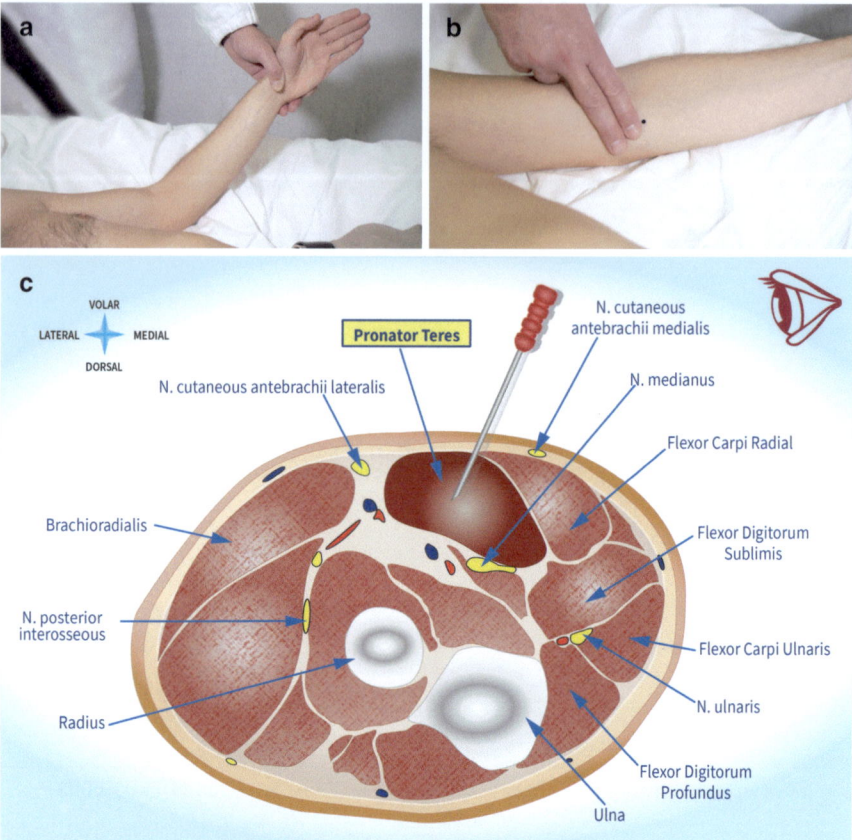

Fig. 11.2 Pronator teres: (**a**) test, (**b**) insertion point, (**c**) cross-section through the upper third of the forearm. The eye shows the perspective of the examiner

Innervation and Nerve Fibers Route
C6, C7, upper and middle trunks, lateral cord, median nerve (Fig. 11.1).

Test and Activation
Ask to pronate the forearm with the elbow partially flexed. Apply pressure at the lower forearm in the direction of supination (Fig. 11.2a).

Needle Insertion
With the subject's forearm fully supinated, insert the needle two fingerbreadths distal to the midpoint of a line connecting the biceps tendon and medial humerus epicondyle (Fig. 11.2b).

Caveat: if the needle is inserted too medially, it will be in the flexor carpi radialis (FCR) or flexor digitorum superficialis (sublimis), both innervated by the median nerve (Fig. 11.2c). If the needle is inserted too deeply, it may reach the median nerve.

Comments

- The pronator teres is the more powerful of the two pronators.
- The pronator teres is the most proximal muscle innervated by the median nerve.
- The pronator teres is a site of entrapment (pronator teres syndrome) as the median nerve enters the forearm between the two heads of the muscle.
- The pronator teres may be, or may not be, involved in pronator teres syndrome, depending on whether the branch that innervates the muscle detaches from the median nerve proximal to or within the muscle itself.
- The pronator teres may be involved in entrapment at the ligament of Struthers.
- The pronator teres is spared in anterior interosseous nerve syndrome.
- The pronator teres may be involved in C6 and C7 radiculopathy.

11.2 Flexor Carpi Radialis (FCR)

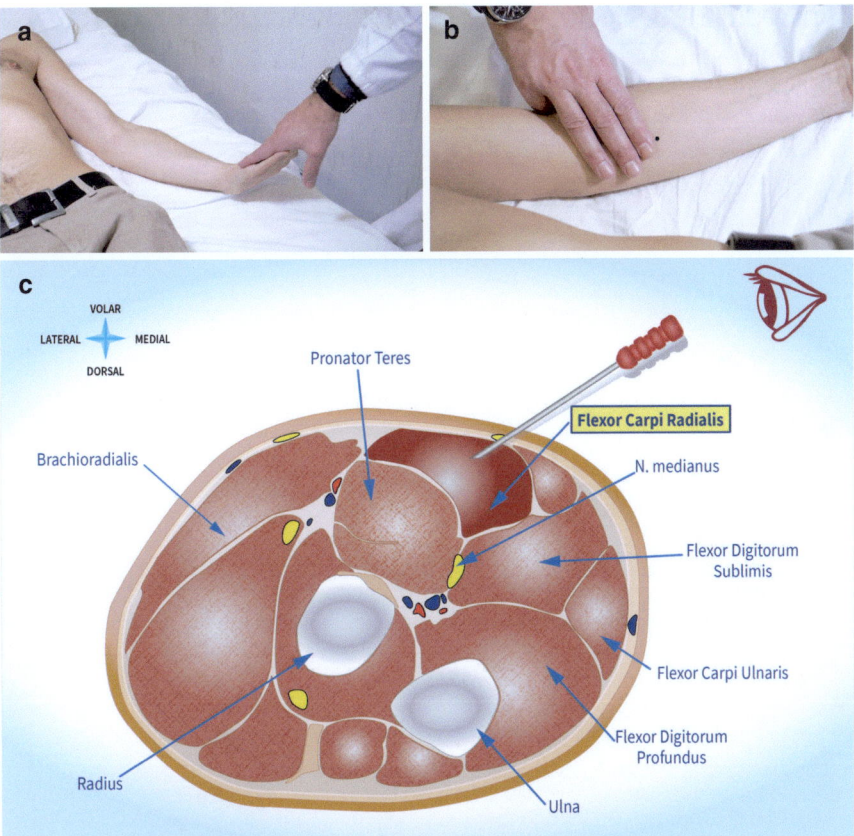

Fig. 11.3 Flexor carpi radialis: (**a**) test, (**b**) insertion point, (**c**) cross-section through the upper third of the forearm. The eye shows the perspective of the examiner

Innervation and Nerve Fibers Route
C6, C7, upper and middle trunks, lateral cord, median nerve (Fig. 11.1).

Test and Activation
With the subject's forearm supinated, ask to flex the wrist toward the radial side. Apply pressure against the thenar eminence in the direction of extension (Fig. 11.3a).

Needle Insertion
Three to four fingerbreadths distal to the midpoint of a line connecting the medial epicondyle and the biceps tendon (Fig. 11.3b).

Caveat: if the needle is inserted too medially, it will be in the flexor digitorum superficialis (sublimis); if inserted too deeply, it may reach the pronator teres and the median nerve (Fig. 11.3c).

Comments

- The FCR is spared in anterior interosseous nerve syndrome.
- The FCR may be affected in proximal median neuropathies including pronator syndrome.
- The FCR may be affected in C6 and C7 radiculopathy.

11.3 Flexor Digitorum Sublimis (FDS)

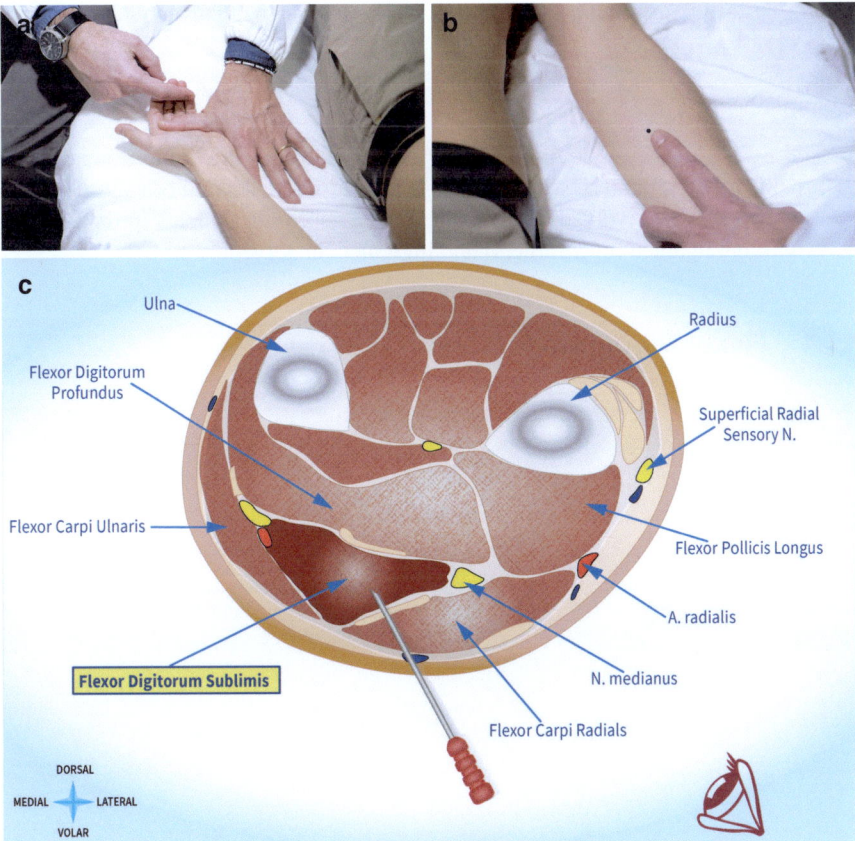

Fig. 11.4 Flexor digitorum sublimis (superficialis): (**a**) test, (**b**) insertion point, (**c**) cross-section through the lower third of the forearm. The eye shows the perspective of the examiner

Innervation and Nerve Fibers Route
C7, C8, middle and lower trunks, lateral and medial cords, median nerve (Fig. 11.1).

Test and Activation
With the hand supinated and stabilizing the metacarpophalangeal joints, ask the subject to flex the digits at the proximal interphalangeal joints. Apply pressure against the palmar surface of the middle phalanx in the direction of extension (Fig. 11.4a).

Needle Insertion
With the subject's forearm supinated, hold the volar surface of the wrist, point the index finger to the biceps tendon, and insert the needle just ulnarly to the tip of the index finger (Fig. 11.4b).

Caveat: if the needle is inserted too laterally, it will be in the flexor carpi radialis, which is also innervated by the median nerve; if inserted too deeply, it will be in the flexor digitorum profundus (FDP), innervated by the median and ulnar nerves (Fig. 11.4c).

Comments

- The flexor digitorum sublimis (FDS) is more difficult to examine than other proximal median muscles, such as the flexor carpi radialis and pronator teres.
- The FDS may be involved in proximal median nerve entrapments (pronator teres syndrome, ligament of Struthers) and high median nerve lesions.

11.4 Flexor Digitorum Profundus (FDP)

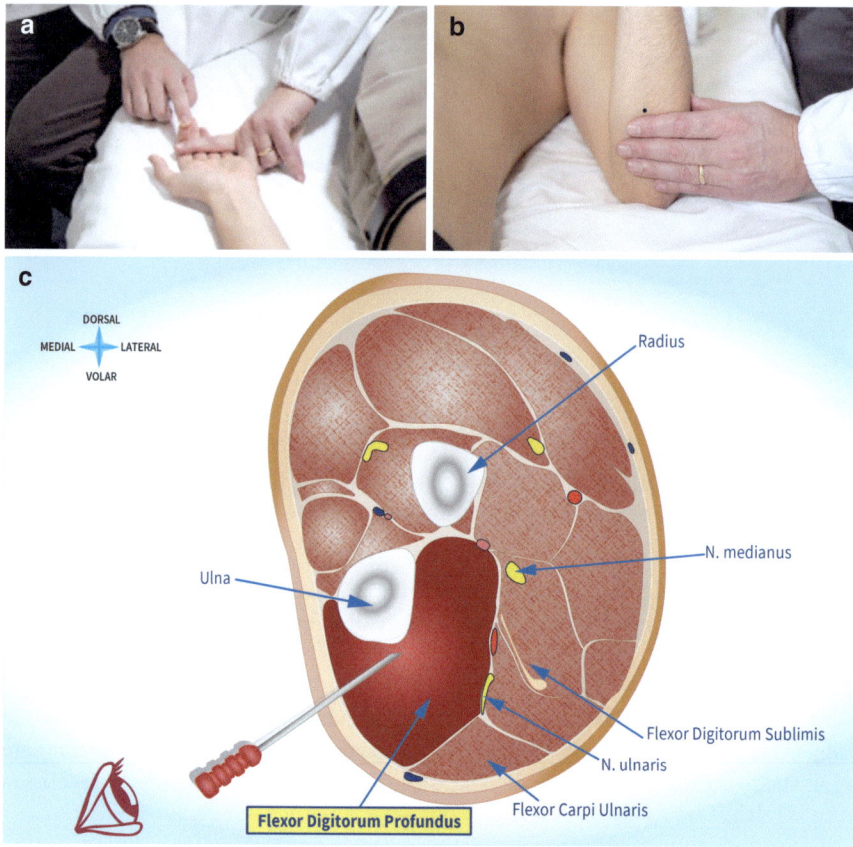

Fig. 11.5 Flexor digitorum profundus: (**a**) test, (**b**) insertion point, (**c**) cross-section through the upper third of the forearm. The eye shows the perspective of the examiner

11.4 Flexor Digitorum Profundus (FDP)

Innervation and Nerve Fibers Route

Digits 2 and 3: C7, C8, T1, middle and lower trunks, medial cord, median nerve, anterior interosseous nerve. Digits 4 and 5: C8, T1, lower trunk, medial cord, ulnar nerve (Fig. 11.1).

Test and Activation

With the hand supinated and stabilizing the proximal and middle phalanges, ask to flex the distal interphalangeal joints. Apply pressure against the palmar surface of the distal phalanx in the direction of extension. Each finger is tested separately, as illustrated in the figure for the index finger (Fig. 11.5a).

Needle Insertion

With the patient's elbow flexed and hand pointing toward the head, insert the needle three to four fingerbreadths distal to the olecranon (Fig. 11.5b). The ulnar-innervated portion is more superficial (1–2 cm), whereas the median-innervated portion is deeper (3–5 cm). The individual muscle portion can be identified by having the subject flex each finger at a time.

Caveat: if the needle is inserted too laterally, it will be in the flexor carpi ulnaris innervated by the ulnar nerve (Fig. 11.5c). When exploring the FPD, to avoid the ulnar nerve, the needle should be angled medially toward the body.

Comments

- The median-innervated portion of the FDP is involved in anterior interosseus nerve lesions, pronator teres, and the ligament of Struthers entrapments.
- Because nerve fascicles can be differentially involved, the ulnar-innervated portion of the FDP is not always involved in ulnar neuropathies at the elbow.

11.5 Flexor Pollicis Longus (FPL)

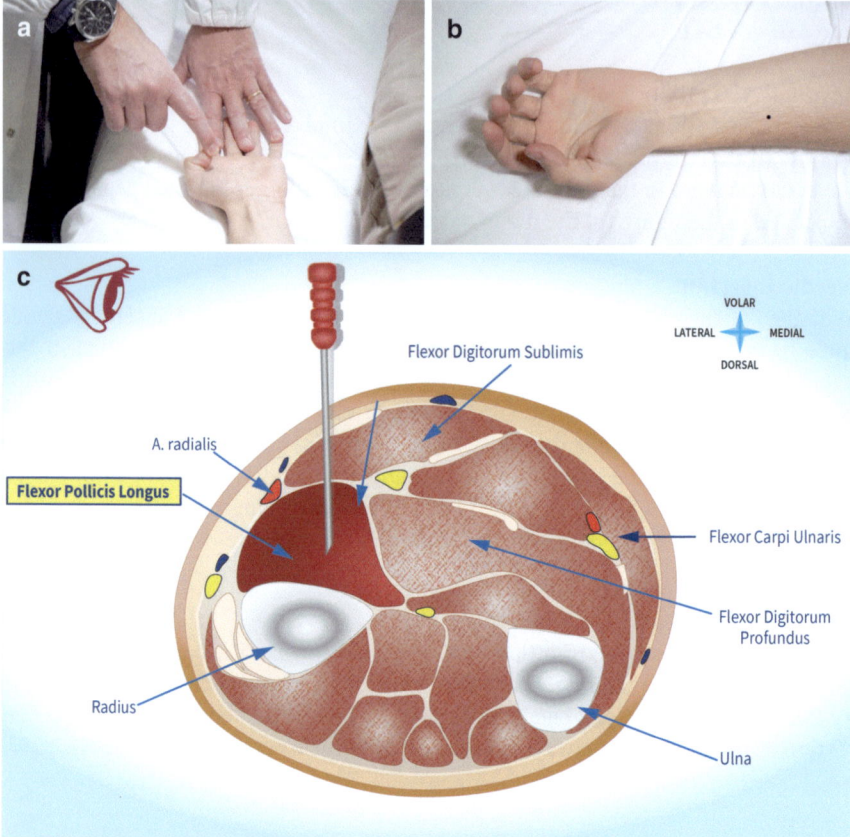

Fig. 11.6 Flexor pollicis longus: (**a**) test, (**b**) insertion point, (**c**) cross-section through the lower third of the forearm. The eye shows the perspective of the examiner

Innervation and Nerve Fibers Route
C7, C8, T1, middle and lower trunks, lateral and medial cords, median nerve, anterior interosseous nerve (Fig. 11.1).

Test and Activation
With the hand supinated, ask to flex the distal phalanx of the thumb. Apply pressure against the palmar surface of the distal phalanx in the direction of extension (Fig. 11.6a).

Needle Insertion
With the forearm supinated, insert the needle at the lower third of a line connecting the lateral wrist to the elbow over the radius (Fig. 11.6b).

Caveat: if the needle is inserted too superficially, it will be in the flexor digitorum superficialis (sublimis), which is also innervated by the median nerve (Fig. 11.6c). The radial artery is just lateral to the insertion point and may be injured.

Comments

- The flexor pollicis longus (FPL) is a distal C8 median-innervated muscle above the wrist.
- The FPL may be affected in anterior interosseus syndrome, pronator teres, and the ligament of Struthers entrapment.
- The FPL is activated by patients with ulnar nerve lesions producing the flexion of the terminal phalanx of the thumb (Froment's sign) to compensate for the paralysis of the adductor pollicis.

11.6 Pronator Quadratus

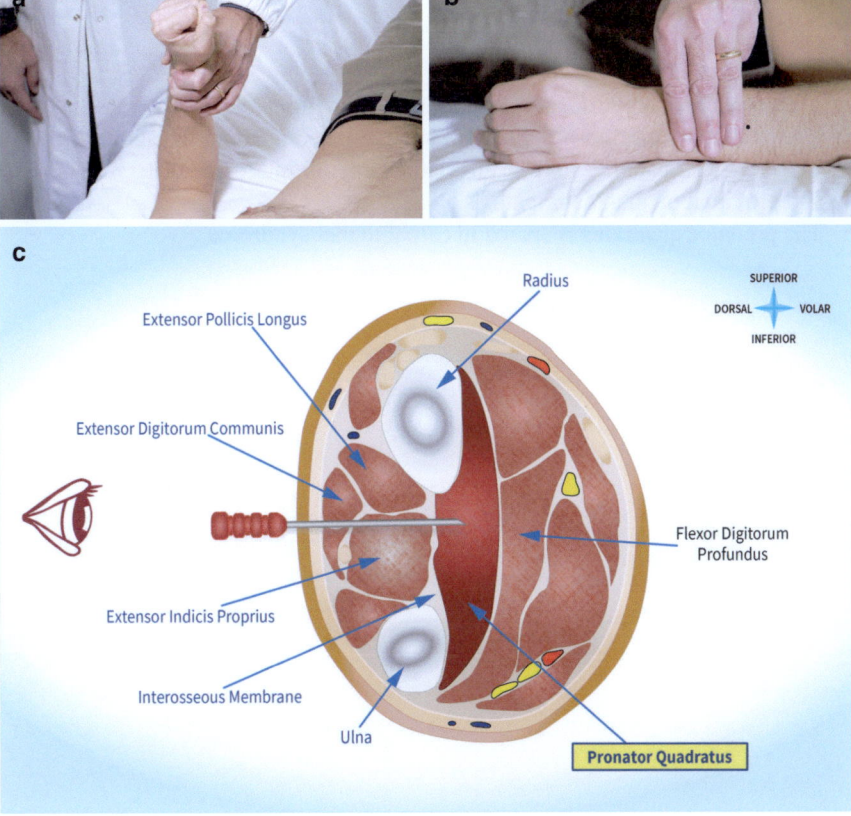

Fig. 11.7 Pronator quadratus: (**a**) test, (**b**) insertion point, (**c**) cross-section through the lower third of the forearm. The eye shows the perspective of the examiner

Innervation and Nerve Fibers Route

C7, C8, T1, middle and lower trunks, lateral and medial cords, median nerve, anterior interosseous nerve (Fig. 11.1).

Test and Activation

Ask to pronate the forearm with the elbow completely flexed to make the action of the pronator teres less effective by being in a shortened position. Apply pressure at the lower forearm-wrist in the direction of forearm supination (Fig. 11.7a).

Needle Insertion

With the forearm in a neutral position, insert the needle three fingerbreadths proximal to the midpoint of a line connecting the radial and ulnar styloids (Fig. 11.7b). Insert the needle through the interosseous membrane at a depth of about 2 cm.

Caveat: if the needle is inserted too superficially, it will be in the extensor digitorum communis, extensor indicis proprius, and extensor pollicis longus; all these muscles are innervated by the radial nerve (Fig. 11.7c). If the needle is inserted too deeply, it will be in the flexor digitorum profundus, innervated by the median and ulnar nerves.

Comments

- The pronator quadratus is the weaker of the two pronators.
- The pronator quadratus is a distal C8 median-innervated muscle above the wrist.
- The pronator quadratus may be affected in anterior interosseous nerve syndrome and proximal median neuropathies.
- The pronator quadratus is spared in carpal tunnel syndrome.

11.7 Abductor Pollicis Brevis (APB)

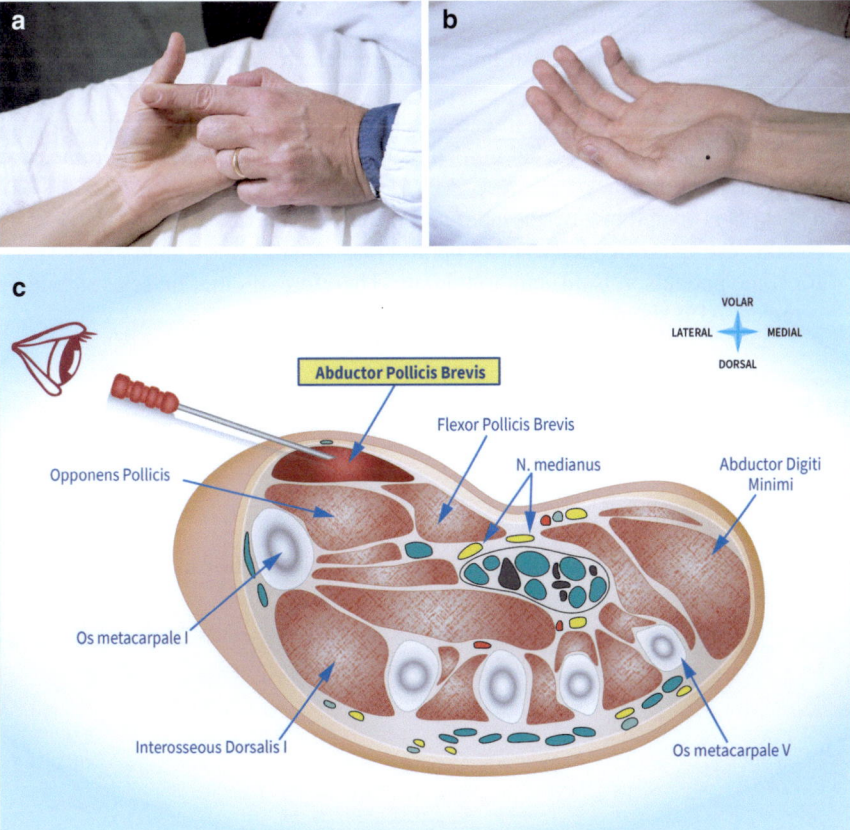

Fig. 11.8 Abductor pollicis brevis: (**a**) test, (**b**) insertion point, (**c**) cross-section through the five metacarpal bones. The eye shows the perspective of the examiner

Innervation and Nerve Fibers Route
C8, T1, lower trunk, medial cord, median nerve (Fig. 11.1).

Test and Activation
With the hand supinated, ask the subject to abduct the thumb from the palm. Apply pressure on the proximal phalanx in the direction of adduction toward the palm (Fig. 11.8a).

Needle Insertion
Tangentially into the thenar eminence at the midpoint of the first metacarpal bone to a depth of 0.5–1 cm (Fig. 11.8b).

Caveat: if the needle is inserted too deeply, it will be in the opponens pollicis, which is also innervated by the median nerve; if the needle is inserted too medially, it will be in the flexor pollicis brevis, which has median and ulnar innervation (Fig. 11.8c).

Comments

- The abductor pollicis brevis (APB) is employed as the recording muscle for median motor conduction studies.
- The APB is the best median-innervated muscle to sample distal to the carpal tunnel.
- The APB may be abnormal in carpal tunnel syndrome, proximal median neuropathies, lower trunk/medial cord plexopathy, thoracic outlet syndrome, and C8–T1 radiculopathies.
- The APB is spared in anterior interosseous nerve syndrome.
- Needle testing is often felt as more painful than other intrinsic hand muscles, and for some patients, it is difficult to tolerate. Therefore, if other muscles should be sampled, APB should not be studied first.

11.8 Opponens Pollicis

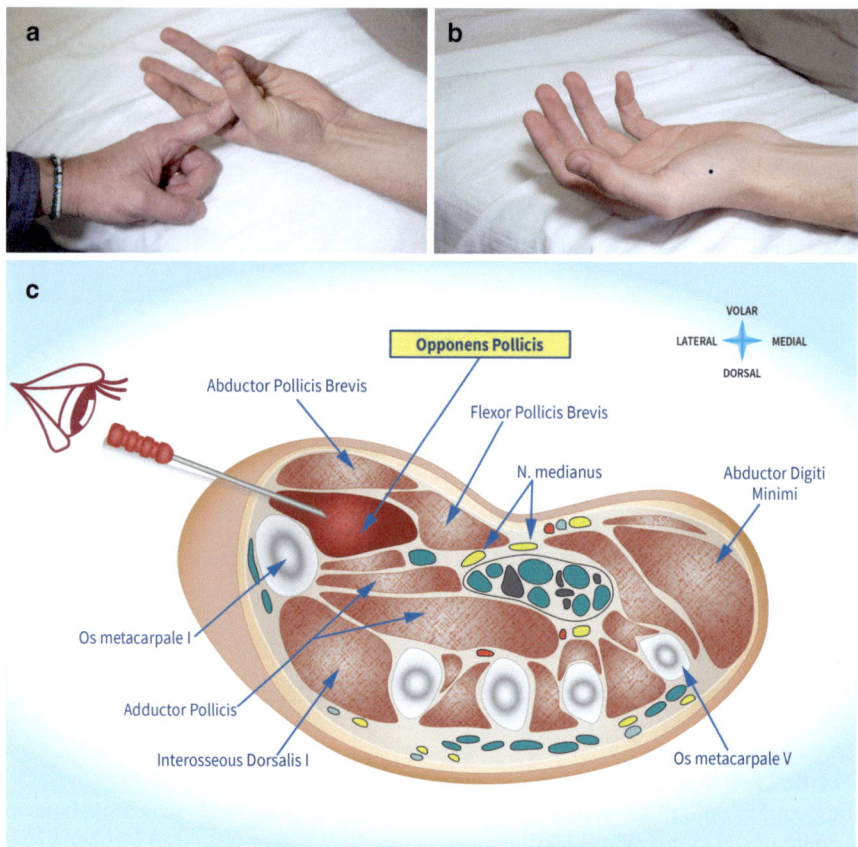

Fig. 11.9 Opponens pollicis: (**a**) test, (**b**) insertion point, (**c**) cross-section through the five metacarpal bones. The eye shows the perspective of the examiner

11.8 Opponens Pollicis

Innervation and Nerve Fibers Route

C8, T1, lower trunk, medial cord, median nerve (Fig. 11.1).

Test and Activation

With the hand supinated, ask the subject to oppose the thumb to the little finger. Apply pressure on the first metacarpal bone in the direction of abduction and extension (Fig. 11.9a).

Needle Insertion

Into the lateral thenar eminence, just above the midpoint of the first metacarpal bone, to a depth of 1.2–1.8 cm (Fig. 11.9b).

Caveat: if the needle is inserted too superficially or medially, it will be in the abductor pollicis brevis, also innervated by the median nerve; if the needle is inserted too deeply, it will be in the adductor pollicis, innervated by the ulnar nerve (Fig. 11.9c).

Comments

- The opponens pollicis may be abnormal in carpal tunnel syndrome, proximal median neuropathies, lower trunk/medial cord plexopathy, thoracic outlet syndrome, and C8–T1 radiculopathies.
- The opponens pollicis is spared in anterior interosseous nerve syndrome.

Muscles Innervated by the Ulnar Nerve

12

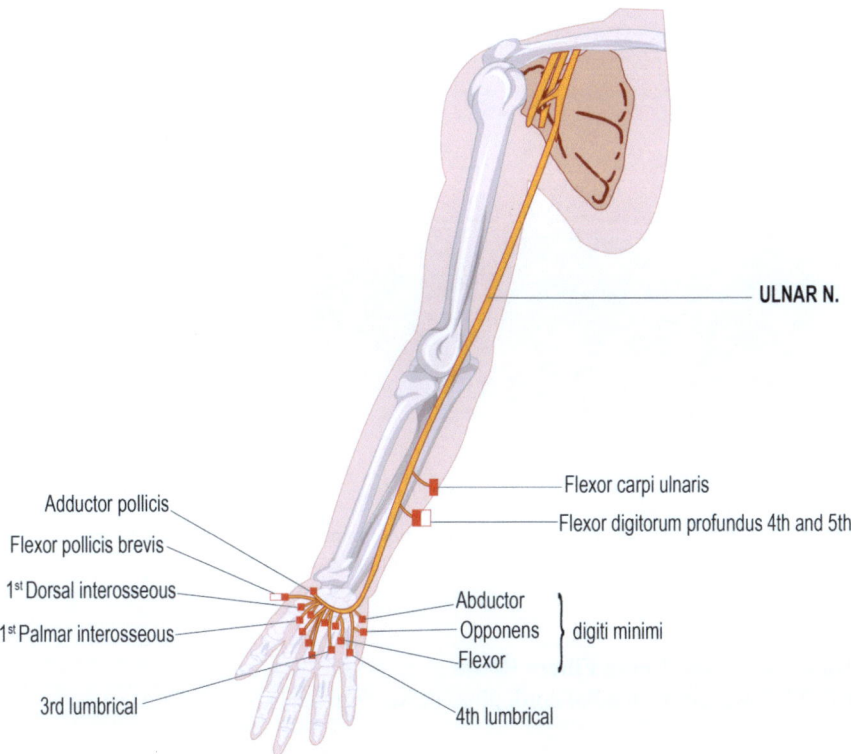

Fig. 12.1 Ulnar nerve and the muscles it supplies. The white part of the label of the flexor digitorum profundus and flexor pollicis brevis indicates that these muscles are innervated also from the median nerve

12.1 Flexor Carpi Ulnaris (FCU)

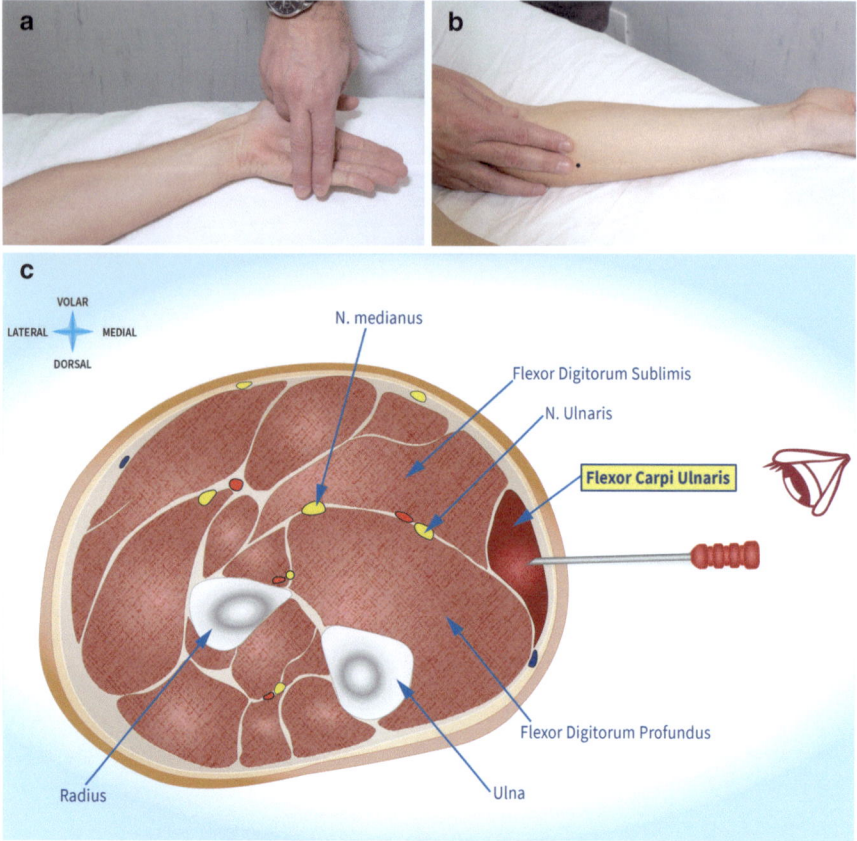

Fig. 12.2 Flexor carpi ulnaris. (**a**) test, (**b**) insertion point, (**c**) cross-section through the upper third of the forearm. The eye shows the perspective of the examiner

Innervation and Nerve Fibers Route
C8, T1, lower trunk, medial cord, ulnar nerve (Fig. 12.1).

Test and Activation
With the subject's forearm in supination, ask to flex the wrist toward the ulnar side. Apply pressure against the hypothenar eminence in the direction of extension toward the radial side (Fig. 12.2a).

Needle Insertion
At the junction of the upper and middle thirds of the forearm, two fingerbreadths medial to the ulna (Fig. 12.2b).

Caveat: if the needle is inserted too deeply, it will be in the flexor digitorum profundus innervated by the median and ulnar nerves (Fig. 12.2c).

Comments

- The flexor carpi ulnaris (FCU) is a very superficial and thin muscle.
- The FCU is not always involved in ulnar neuropathies at the elbow because the nerve branch to the FCU occasionally arises proximal to the medial epicondyle, or the ulnar nerve fascicles can be differentially affected.

12.2 Abductor Digiti Minimi (ADM)

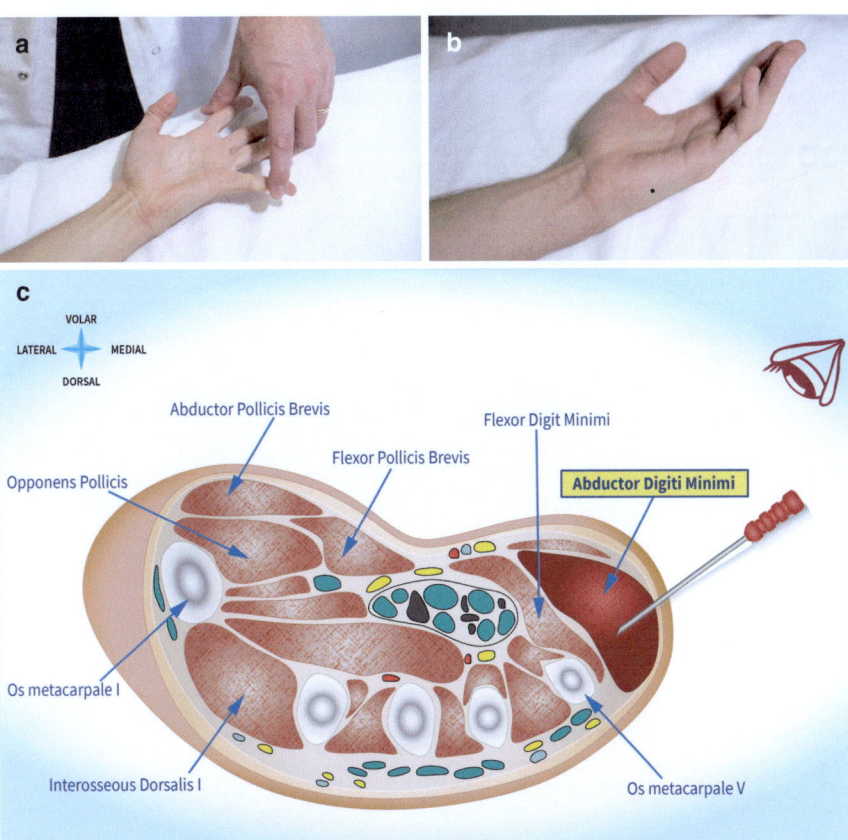

Fig. 12.3 Abductor digiti minimi: (**a**) test, (**b**) insertion point, (**c**) cross-section anatomy through the five metacarpal bones. The eye shows the perspective of the examiner

Innervation and Nerve Fibers Route
C8, T1, lower trunk, medial cord, ulnar nerve (Fig. 12.1).

Test and Activation
With the hand supinated, ask the subject to abduct the little finger. Apply pressure against the ulnar side of the little finger (Fig. 12.3a).

Needle Insertion

At the midpoint of the fifth metacarpal bone (Fig. 12.3b).

Caveat: if inserted too deeply, the needle will be in the flexor or opponens digiti minimi, both innervated by the ulnar nerve (Fig. 12.3c).

Comments

- The abductor digiti minimi (ADM) is used as the recording muscle for the ulnar nerve motor conduction study.
- The ADM may be involved in ulnar neuropathy, lower trunk/medial cord plexopathy, thoracic outlet syndrome, and C8 and T1 radiculopathies.
- The ADM is spared in ulnar lesions at the distal Guyon's canal.

12.3 Interosseous Dorsalis I

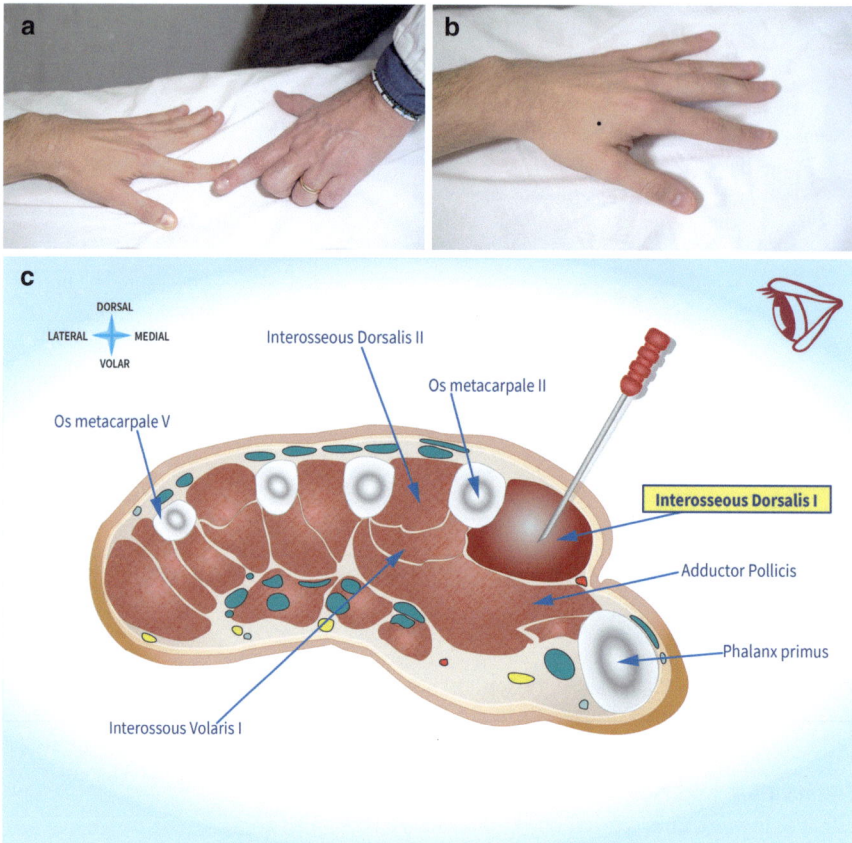

Fig. 12.4 Interosseous dorsalis I: (**a**) test, (**b**) insertion point, (**c**) cross-section through the first phalanx of the thumb and the II, III, IV, V metacarpal bones. The eye shows the perspective of the examiner

12.3 Interosseous Dorsalis I

Innervation and Nerve Fibers Route
C8, T1, lower trunk, medial cord, ulnar nerve, deep branch of ulnar nerve (Fig. 12.1).

Test and Activation
With the palm and fingers flat over a surface, ask to abduct the index finger toward the radial side. Apply pressure against the distal phalanx (Fig. 12.4a).

Needle Insertion
Just radial to the second metacarpal bone (Fig. 12.4b).

Caveat: if the needle is inserted too deeply, it will be in the adductor pollicis that is also innervated by the ulnar nerve (Fig. 12.4c).

Comments

- The interosseous dorsalis I is employed as the recording muscle in motor conduction studies of the deep branch of the ulnar nerve.
- The interosseous dorsalis I may be involved in ulnar lesions at Guyon's canal, in ulnar neuropathy, lower trunk/medial cord plexopathy, thoracic outlet syndrome, and C8 and T1 radiculopathies.
- The interosseous dorsalis I is affected, and the ADM is normal at distal Guyon's canal entrapment because of the involvement of only the ulnar deep motor branch.
- The interosseous dorsalis I is the least painful of the intrinsic hand muscles at needle testing.

12.4 Flexor Pollicis Brevis (FPB)

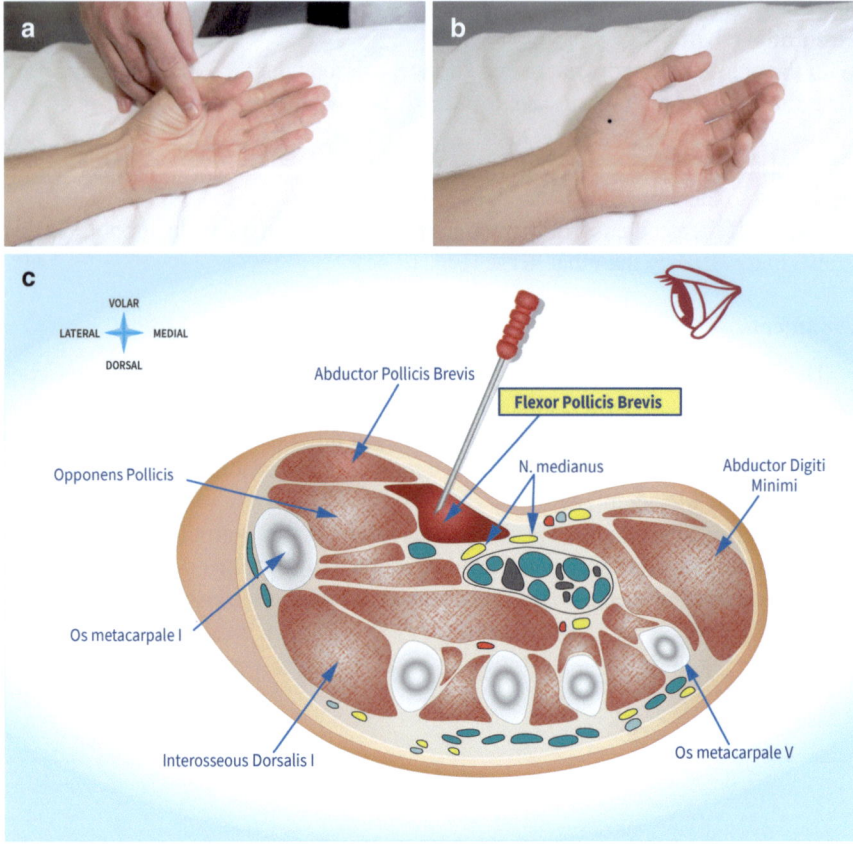

Fig. 12.5 Flexor pollicis brevis: (**a**) test, (**b**) insertion point, (**c**) cross-section through the five metacarpal bones. The eye shows the perspective of the examiner

Innervation and Nerve Fibers Route
Superficial head: C8, T1, lower trunk, medial cord, median nerve. Deep head: C8, T1, lower trunk, medial cord, ulnar nerve (Fig. 12.1).

Test and Activation
With the hand supinated, ask to flex the thumb at the metacarpophalangeal joint. Apply pressure against the palmar surface of the proximal phalanx in the direction of extension (Fig. 12.5a).

12.4 Flexor Pollicis Brevis (FPB)

Needle Insertion

In the thenar eminence just medial to the midpoint of the first metacarpal bone (Fig. 12.5b). The superficial head is about 0.5–1 cm deep, and the deep head is about 1.5–2 cm deep.

Caveat: if the needle is inserted too laterally, it will be in the abductor pollicis brevis; if the needle is inserted too deeply, it will be in the opponens pollicis, both innervated by the median nerve (Fig. 12.5c).

Comments

- The superficial head of flexor pollicis brevis (FPB) is usually median innervated, while the deep head is ulnar innervated. However, innervation varies widely, and in some subjects, both heads are median or ulnar innervated.
- Because of normal anatomic variation, abnormalities should be interpreted with caution when trying to separate median from ulnar lesions.

Muscles Innervated by the Femoral and Obturator Nerves

13

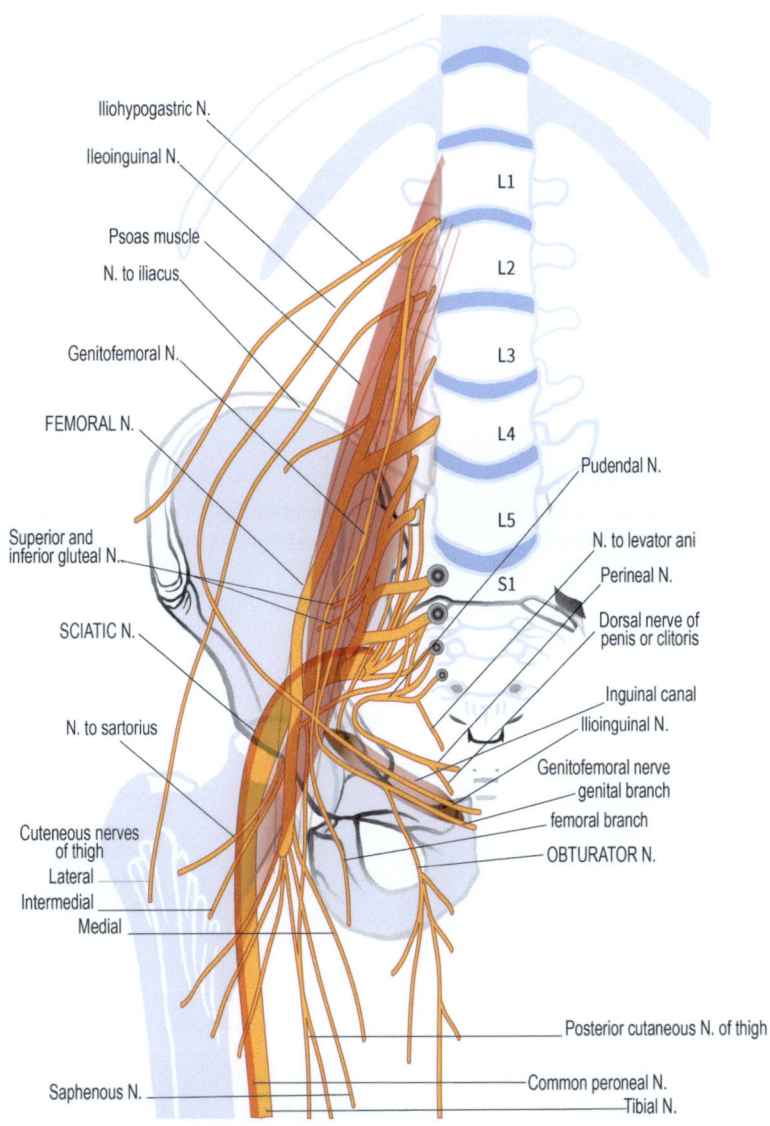

Fig. 13.1 The lumbosacral plexus and its branches

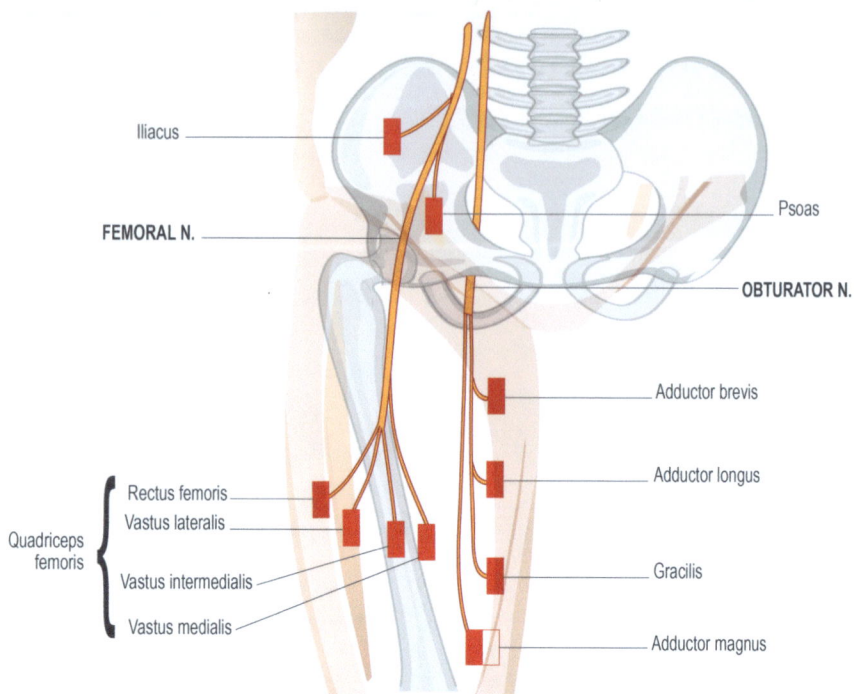

Fig. 13.2 The femoral and obturator nerves and the muscles which they supply. The white part of the label of the adductor magnus indicates that this muscle is innervated also from the sciatic nerve

13.1 Femoral Nerve

13.1.1 Iliacus

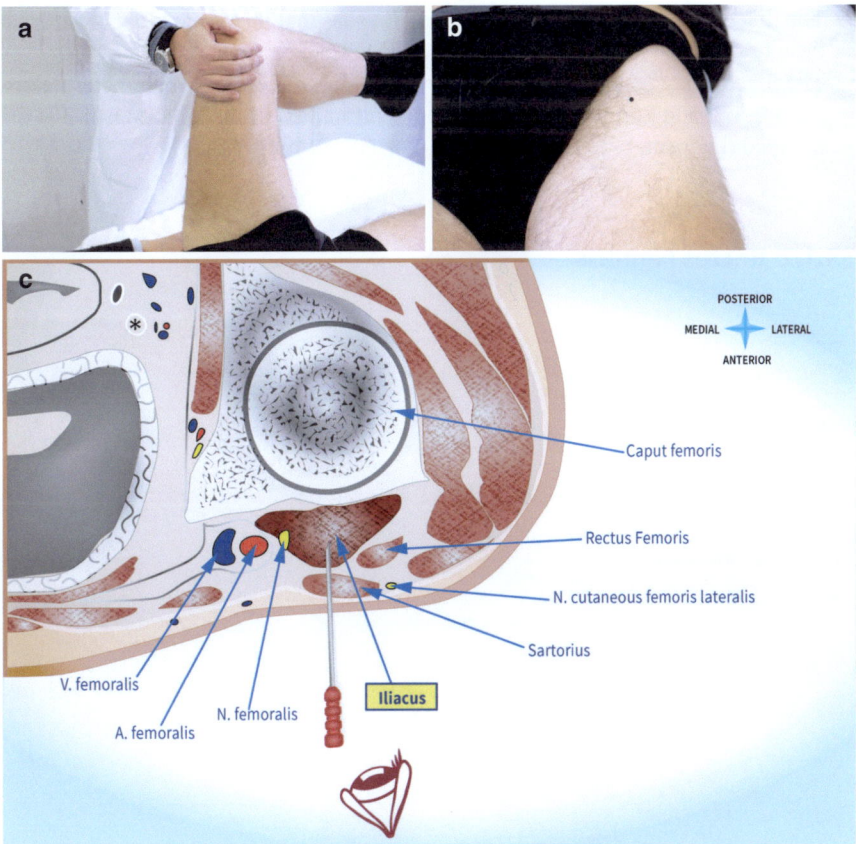

Fig. 13.3 Iliacus: (**a**) test, (**b**) insertion point, (**c**) cross-section through the acetabular fossa and the head of the femur. The eye shows the perspective of the examiner

Innervation and Nerve Fibers Route
L2, L3, L4, lumbar plexus, femoral nerve (Figs. 13.1 and 13.2).

Test and Activation
With the subject lying supine, ask to flex the thigh with the knee flexed (Fig. 13.3a).

Needle Insertion
Two fingerbreadths lateral to the pulse of the femoral artery and one fingerbreadth below the inguinal ligament (Fig. 13.3b).

Caveat: if the needle is inserted too superficially and laterally, it will be in the sartorius; if the needle is inserted too medially, it may injure the neurovascular bundle (femoral artery, femoral vein, and femoral nerve) (Fig. 13.3c). The lateral femoral cutaneous nerve is just lateral to the insertion point.

Comments

- The iliacus and psoas muscles form the iliopsoas that flexes the hip. However, at the site shown in the Fig. 13.3, only the iliacus can be sampled.
- The iliacus may be involved in lesions of the high femoral nerve, lumbar plexus, and L2, L3, and L4 roots, but is spared in entrapment of the femoral nerve at the inguinal ligament.

13.1.2 Rectus Femoris

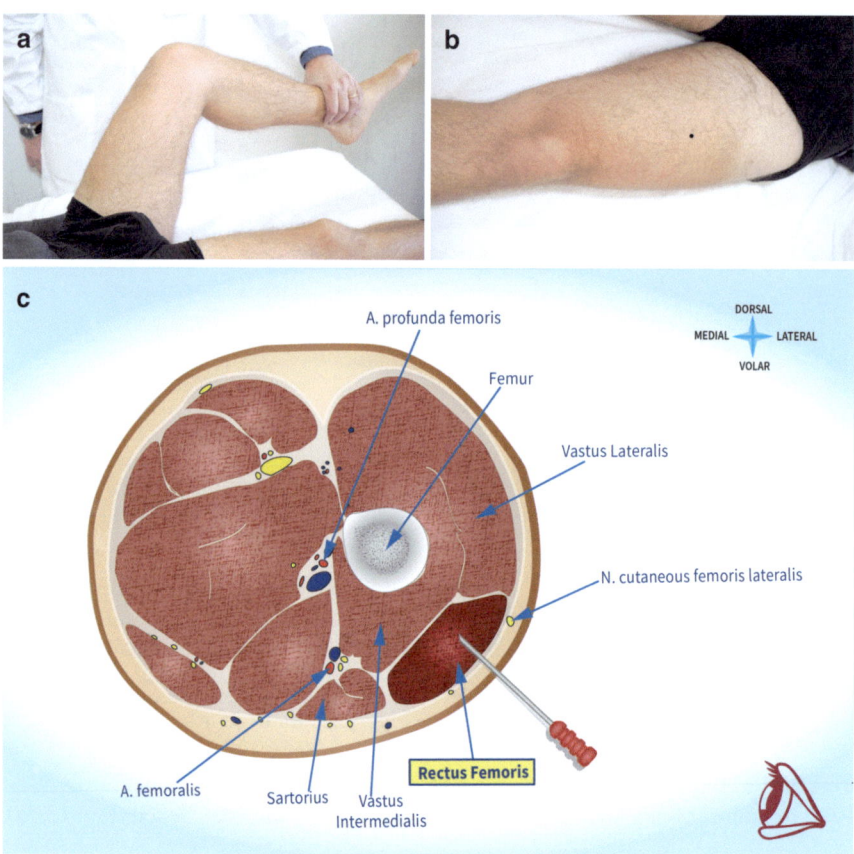

Fig. 13.4 Rectus femoris: (**a**) test, (**b**) insertion point, (**c**) cross-section through the upper third of the thigh. The eye shows the perspective of the examiner

13.1 Femoral Nerve

Innervation and Nerve Fibers Route

L2, L3, L4, lumbar plexus, femoral nerve (Figs. 13.1 and 13.2).

Test and Activation

With the subject lying supine, ask to extend the knee while lifting the heel from the bed (Fig. 13.4a).

Needle Insertion

On the anterior aspect of the thigh, midway between the superior border of the patella and the anterior superior iliac spine (Fig. 13.4b).

Caveat: if the electrode is inserted too medially and too deeply, it will be in the vastus intermedius; if the needle is inserted too laterally, it will be in the vastus lateralis; if inserted too distally and medially, it will be in the vastus medialis (Fig. 13.4c). However, all these muscles are innervated by the same nerve (femoral) and roots (L2, L3, L4).

Comments

- The rectus femoris is more a flexor of the hip than a knee extensor.
- The rectus femoris is more difficult to activate than the vastus lateralis.
- The rectus femoris may be involved in lesions of the femoral nerve at or proximal to the inguinal ligament, lumbar plexus, and L2, L3, and L4 roots.

13.1.3 Vastus Lateralis

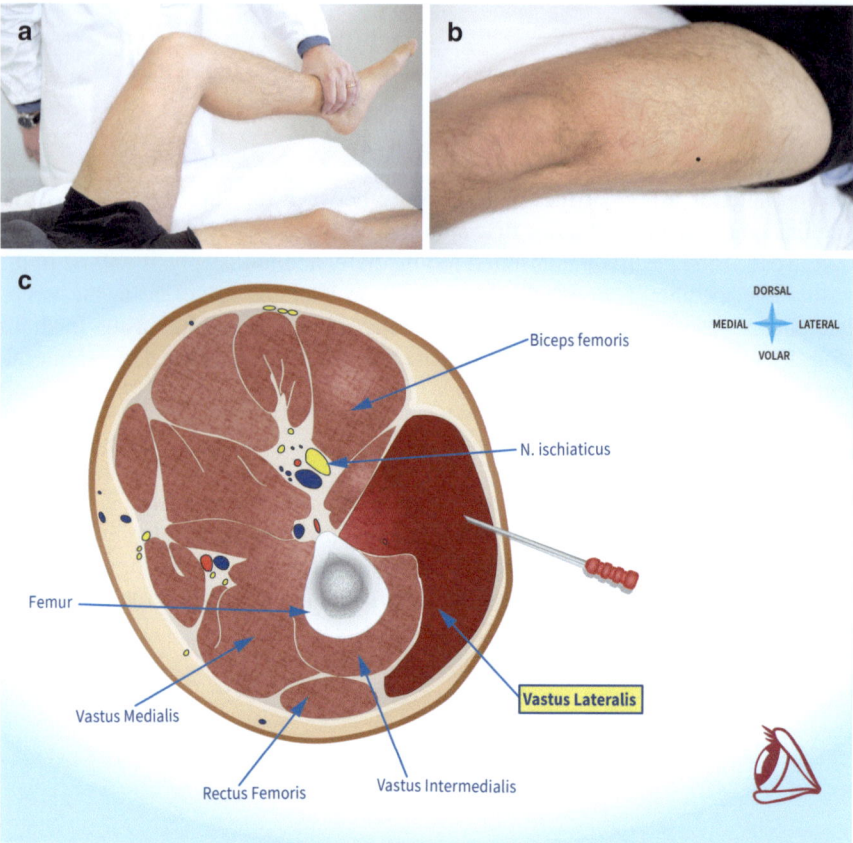

Fig. 13.5 Vastus lateralis: (**a**) test, (**b**) insertion point, (**c**) cross-section through the middle of the thigh. The eye shows the perspective of the examiner

Innervation and Nerve Fibers Route
L2, L3, L4, lumbar plexus, femoral nerve (Figs. 13.1 and 13.2).

Test and Activation
With the subject lying supine, ask to extend the knee while lifting the heel from the bed (Fig. 13.5a).

Needle Insertion

Into the lateral thigh four to five fingerbreadths proximal to the patella (Fig. 13.5b).

Caveat: if inserted too medially, the needle will be in the rectus femoris; if inserted too deeply, the needle will be in the vastus intermedius, both innervated by the same nerve (femoral) and roots (L2, L3, L4) (Fig. 13.5c). If the needle is inserted too posteriorly, it will be in the biceps femoris innervated by the sciatic nerve.

Comments

- The vastus lateralis is the head of the quadriceps femoris usually assayed.
- The vastus lateralis may be involved in lesions of the femoral nerve at or proximal to the inguinal ligament, lumbar plexus, and L2, L3, and L4 roots.

13.2 Obturator Nerve, Adductor Magnus

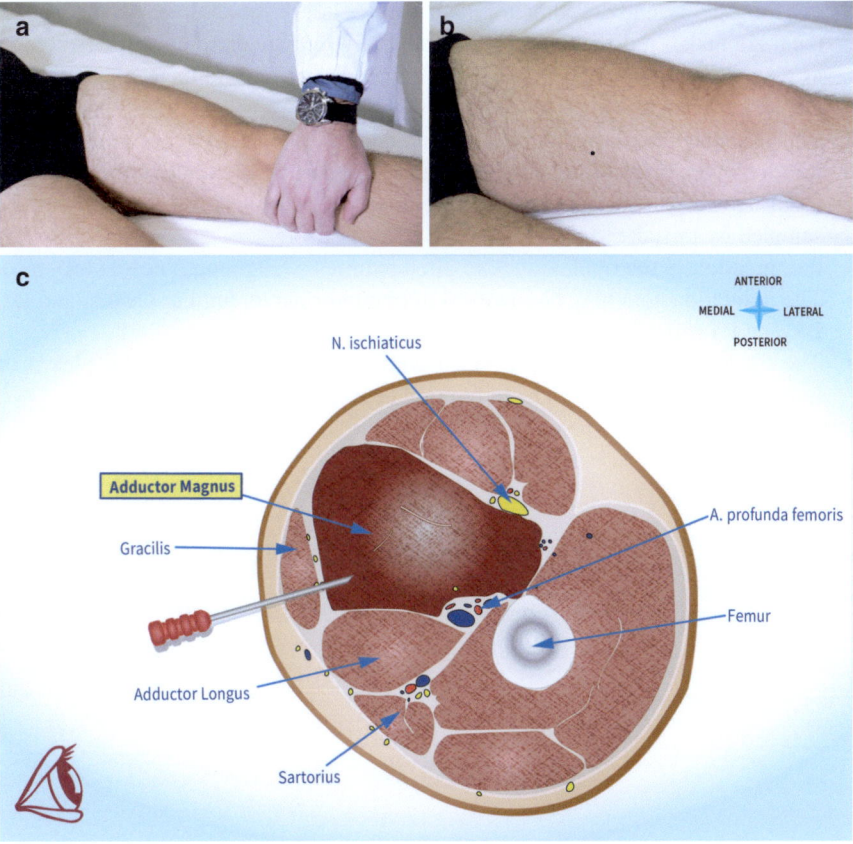

Fig. 13.6 Adductor magnus: (**a**) test, (**b**) insertion point, (**c**) cross-section anatomy through the middle of the thigh. The eye shows the perspective of the examiner

Innervation and Nerve Fibers Route
L2, L3, L4, lumbar plexus, obturator nerve. A small branch comes from the sciatic nerve (Figs. 13.1 and 13.2).

Test and Activation
With the subject lying supine with the lower extremities slightly abducted and externally rotated, ask to adduct the thigh (Fig. 13.6a).

Needle Insertion
Midway between the medial femoral epicondyle and the pubis (Fig. 13.6b).

Caveat: if the needle is inserted too superficially, it will be in the gracilis that is also innervated by the obturator nerve; if the needle is inserted too laterally, it will be in the sartorius innervated by the femoral nerve (Fig. 13.6c).

Comments

- The tight adductors are a functional unit including the adductor longus and brevis, the gracilis, and the adductor magnus muscles. In some subjects, adiposity of the thigh makes it difficult to anatomically differentiate these muscles. However, as they are all innervated by the same nerve (obturator) and the same roots (L2, L3, L4), it is not essential to sample a specific one.
- The deep and lateral part of the adductor magnus is supplied by the sciatic nerve; it is considered part of the hamstring muscles and has the function to extend the thigh. However, this part of the muscle is deep, and it is difficult for it to be sampled by mistake.
- Examination of the tight adductors often requires a long length needle (50 mm).
- The tight adductors may be involved in obturator nerve, lumbar plexus, and L2, L3, and L4 root injuries.
- The tight adductors are useful to differentiate lesions of the lumbar plexus or lumbar roots (are affected) from femoral neuropathy (are spared).

Muscles Innervated by the Superior and Inferior Gluteal Nerves

14

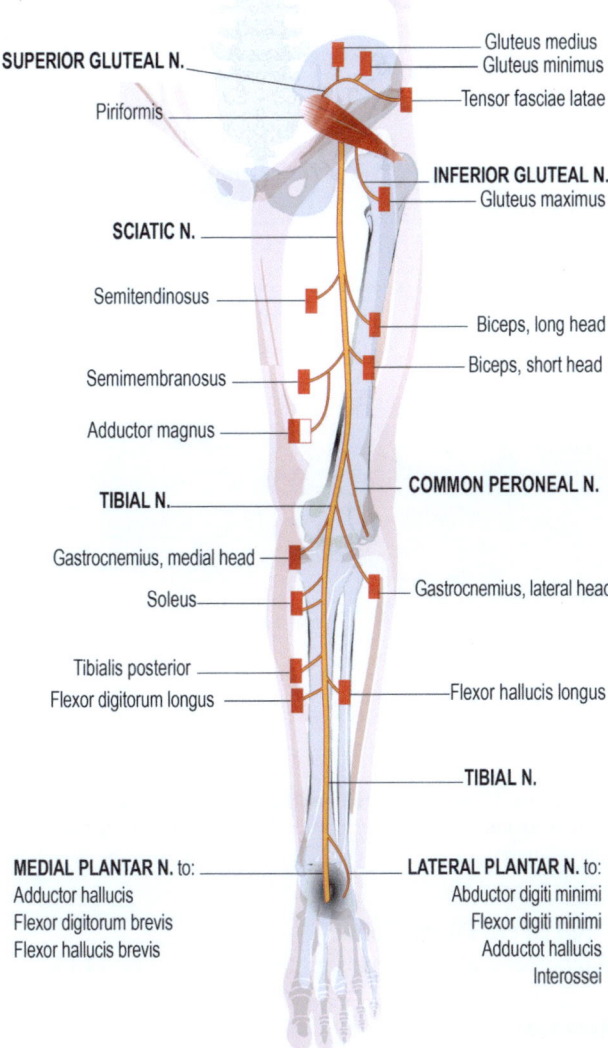

Fig. 14.1 Superior and inferior gluteal, sciatic, and tibial nerves and the muscles which they supply. The white part of the label of the adductor magnus indicates that this muscle is innervated also from the obturator nerve. The short head of the biceps femoris is innervated by the peroneal division of the sciatic nerve

14.1 Superior Gluteal Nerve

14.1.1 Tensor Fasciae Latae (TFL)

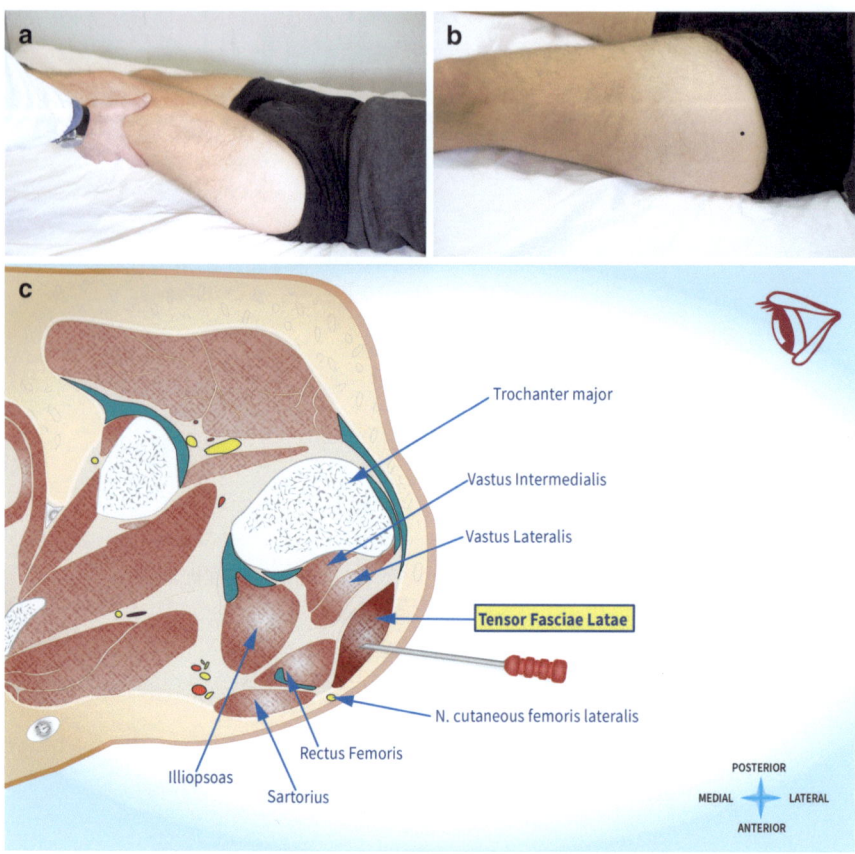

Fig. 14.2 Tensor fasciae latae: (**a**) test, (**b**) insertion point, (**c**) cross-section through the upper portion of tuber ischiaticum and the lower portion of the greater trochanter. The eye shows the perspective of the examiner

Innervation and Nerve Fibers Route
L4, L5, S1, lumbosacral plexus, superior gluteal nerve (Fig. 14.1).

Test and Activation
With the subject lying supine, ask to abduct the thigh with the hip flexed and knee extended. Apply pressure in the direction of adduction (Fig. 14.2a).

14.1 Superior Gluteal Nerve

Needle Insertion

Two fingerbreadths distal to the greater trochanter. The muscle is very superficial (Fig. 14.2b).

Caveat: if the needle is inserted too anteriorly, it will be in the sartorius or rectus femoris; if the needle is inserted too deeply, it will be in the vastus lateralis (Fig. 14.2c). All these muscles are innervated by the femoral nerve. The lateral femoral cutaneous nerve is medial to the insertion site.

Comments

- The tensor fasciae latae (TFL) is a proximal predominantly L5-innervated muscle.
- The TFL is employed to differentiate lesions of the lumbosacral plexus or L5 and S1 roots (it is affected) from sciatic nerve lesions (it is spared).

14.1.2 Gluteus Medius

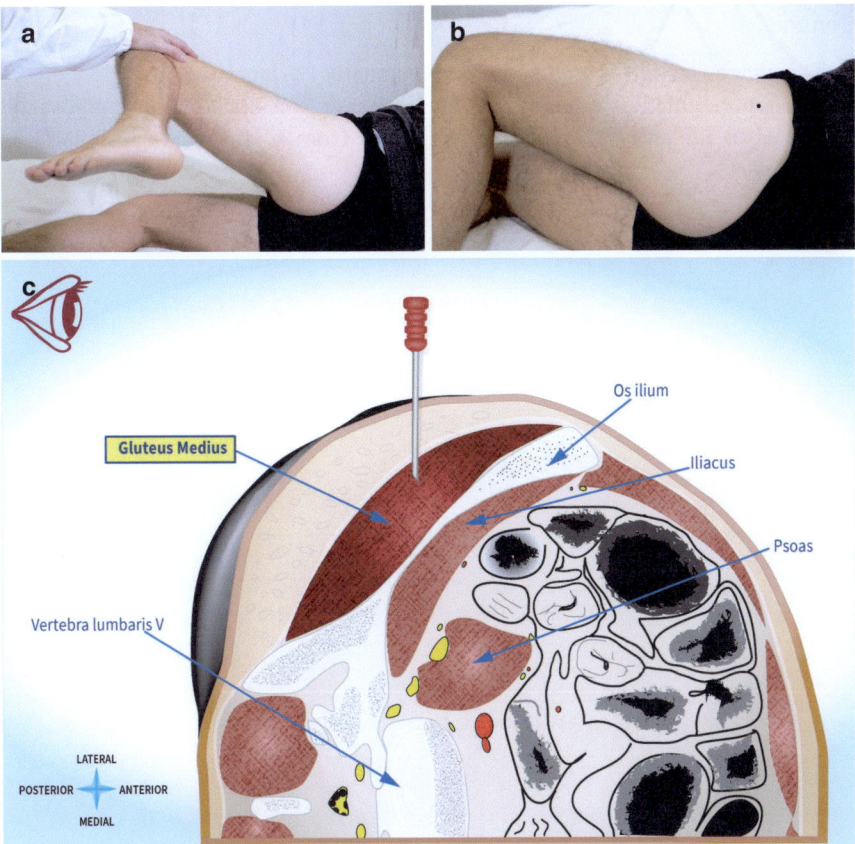

Fig. 14.3 Gluteus medius: (**a**) test, (**b**) insertion point, (**c**) cross-section through the lower margin of the fifth lumbar vertebra and the iliac bone about 5 mm below the superior margin. The eye shows the perspective of the examiner

Innervation and Nerve Fibers Route
L4, L5, S1, lumbosacral plexus, superior gluteal nerve (Fig. 14.1).

Test and Activation
With the subject lying on his/her side and the side to be tested placed upward, ask to abduct the thigh. Apply pressure against the knee in the direction of adduction (Fig. 14.3a).

Needle Insertion
With the subject lying on his/her side and the side to be studied placed upward, insert the needle into the lateral thigh two to three fingerbreadths distal to the iliac crest (Fig. 14.3b).

Caveat: if the needle is inserted too anteriorly, it will be in the tensor fasciae latae also innervated by the superior gluteal nerve; if the electrode is inserted too posteriorly, it will be in the gluteus maximus innervated by the inferior gluteal nerve (Fig. 14.3c).

Comments

- The gluteus medius is a proximal predominantly L5-innervated muscle.
- The gluteus medius is employed to differentiate lesions of the lumbosacral plexus or L5 and S1 roots (it is affected) from sciatic nerve lesions (it is spared).

14.2 Inferior Gluteal Nerve, Gluteus Maximus

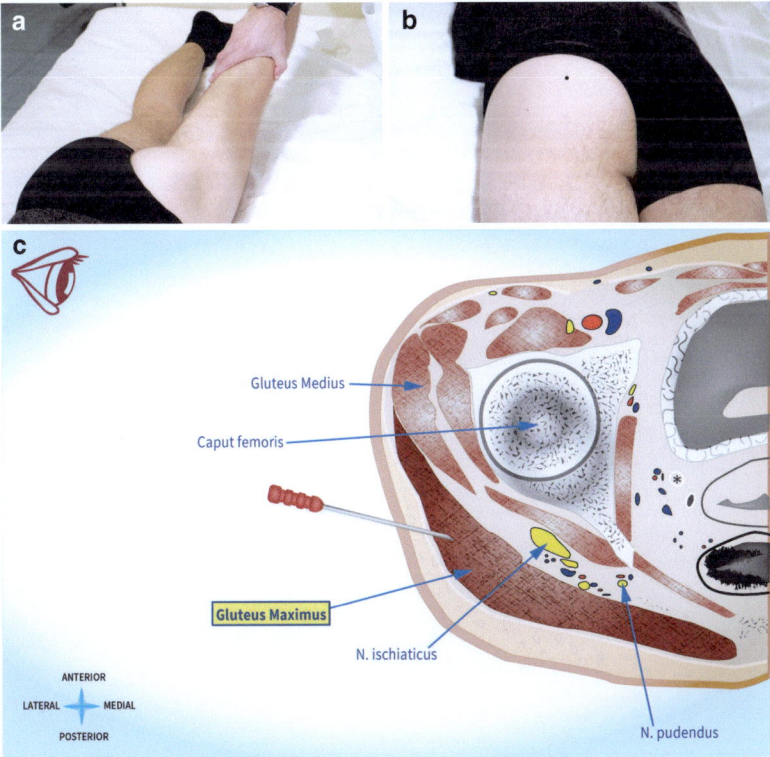

Fig. 14.4 Gluteus maximus: (**a**) test, (**b**) insertion point, (**c**) cross-section through the acetabular fossa and the head of the femur. The eye shows the perspective of the examiner

Innervation and Nerve Fibers Route
L5, S1, S2, lumbosacral plexus, inferior gluteal nerve (Fig. 14.1).

Test and Activation
With the subject prone, ask to extend the thigh with the knee straight (Fig. 14.4a).

Needle Insertion
Into the upper outer quadrant of the buttock (Fig. 14.4b).

Caveat: if the needle is inserted too deeply in the center or in the lower outer quadrant of the buttock, the sciatic nerve can be injured (Fig. 14.4c).

Comments
- Gluteus maximus is the best proximal S1-innervated muscle to test for S1 radiculopathy.
- Gluteus maximus is employed to differentiate lesions of the lumbosacral plexus or L5 and S1 roots (it is affected) from sciatic nerve lesions (it is spared).

Muscles Innervated by the Sciatic and Tibial Nerves

15

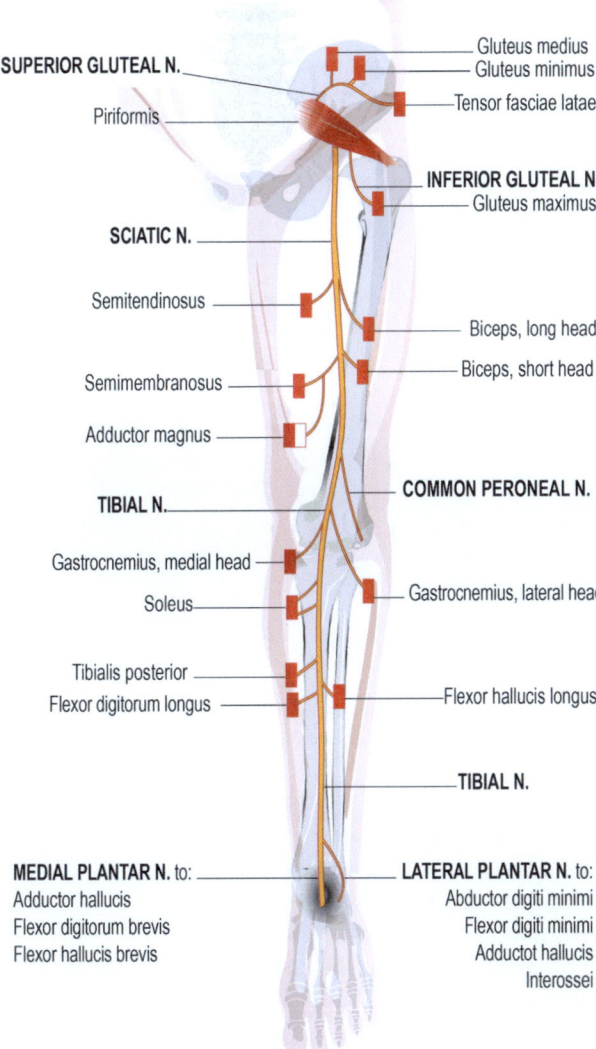

Fig. 14.1 Superior and inferior gluteal, sciatic, and tibial nerves and the muscles which they supply. The white part of the label of the adductor magnus indicates that this muscle is innervated also from the obturator nerve. The short head of the biceps femoris is innervated by the peroneal division of the sciatic nerve

15.1 Biceps Femoris-Caput Longum (BF-CL)

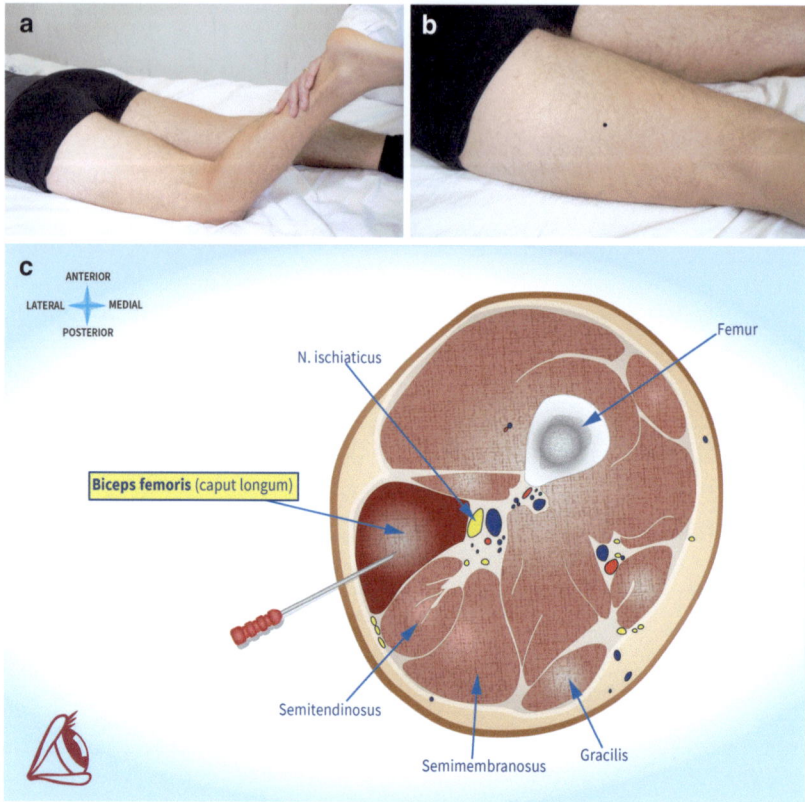

Fig. 15.1 Biceps femoris-caput longum: (**a**) test, (**b**) insertion point, (**c**) cross-section through the middle of the thigh. The eye shows the perspective of the examiner

Innervation and Nerve Fibers Route
L5, S1, lumbosacral plexus, sciatic nerve (tibial division) (Fig. 14.1).

Test and Activation
With the subject prone, ask to flex the knee. Apply pressure against the leg in the direction of knee extension (Fig. 15.1a).

Needle Insertion
At the midpoint of a line between the fibular head and the ischial tuberosity (Fig. 15.1b).

Caveat: if the needle is inserted too medially, it will be in the semitendinosus, innervated by the tibial nerve (Fig. 15.1c).

Comments

- Usually the lateral hamstrings (biceps femoris short and long head) are predominantly S1 innervated, whereas the medial hamstrings (semitendinosus and semimembranosus) are L5 innervated.
- The biceps femoris-caput longum (BF-CL) may be involved in sciatic nerve, lumbosacral plexus, and L5 and S1 lesions.

15.2 Semimembranosus

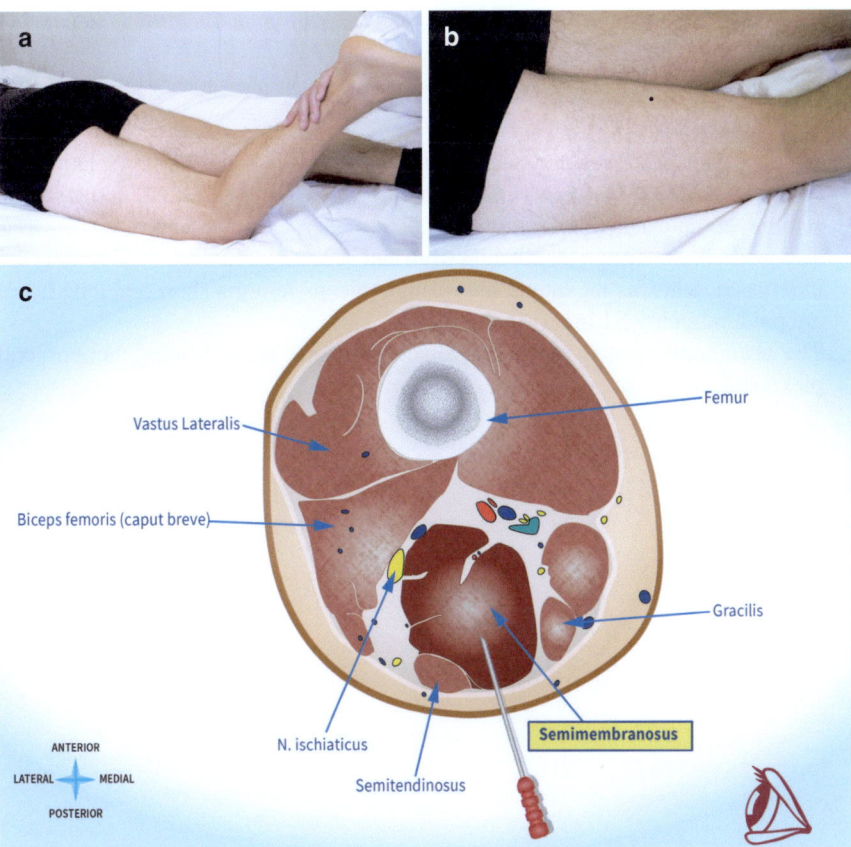

Fig. 15.2 Semimembranosus: (**a**) test, (**b**) insertion point, (**c**) cross-section through the lower third of the thigh. The eye shows the perspective of the examiner

Innervation and Nerve Fibers Route
L4, L5, S1, lumbosacral plexus, sciatic nerve (tibial division) (Fig. 14.1).

Test and Activation
With the subject prone, ask to flex the knee with medial rotation of the tibia. Apply pressure against the leg in the direction of knee extension (Fig. 15.2a).

Needle Insertion
Three to four fingerbreadths above the medial knee (Fig. 15.2b). At this location, the semitendinosus is predominantly tendinous, and the semimembranosus can be easily tested.

Caveat: if the needle is inserted too medially, it will be in the gracilis innervated by the obturator (Fig. 15.2c).

Comments

- The semimembranosus and semitendinosus have the same nerve (sciatic) and root (L4, L5, S1) innervation; therefore, it is suggested to test the semimembranosus.
- The medial hamstrings (semitendinosus and semimembranosus) are usually L5 innervated, whereas the lateral hamstrings (biceps femoris short and long head) are predominantly S1 innervated.
- The semimembranosus may be involved in sciatic nerve lesions, lumbosacral plexopathy, or L5 radiculopathy.

15.3 Gastrocnemius-Caput Mediale (GCM)

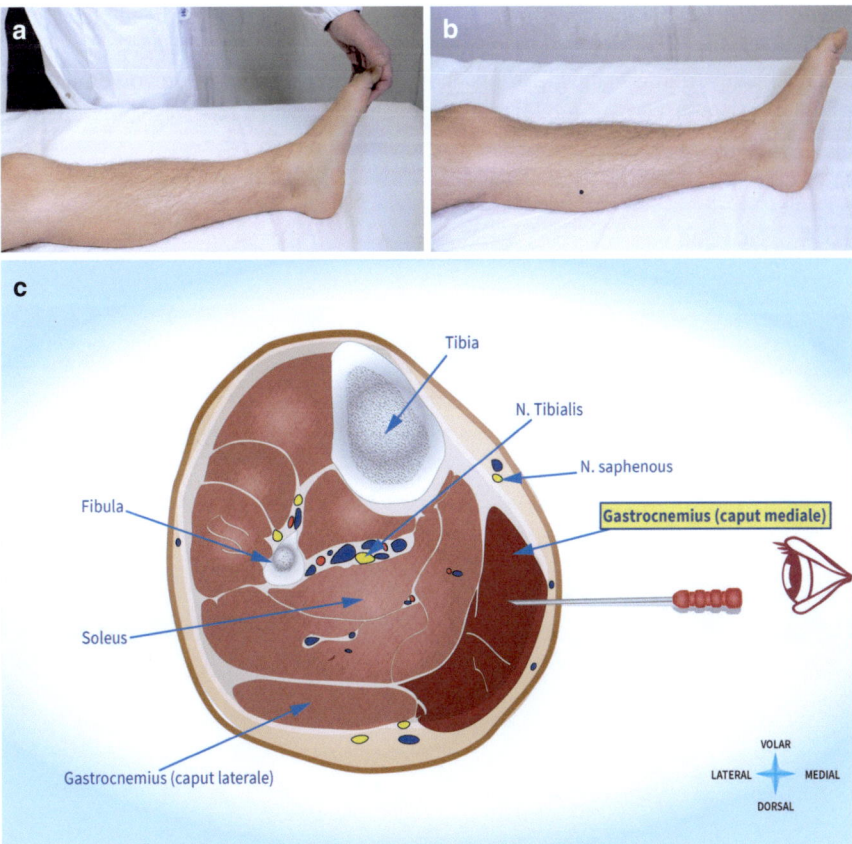

Fig. 15.3 Gastrocnemius-caput mediale: (**a**) test, (**b**) insertion point, (**c**) cross-section through the upper third of the leg. The eye shows the perspective of the examiner

Innervation and Nerve Fibers Route
S1, S2, lumbosacral plexus, sciatic nerve, tibial nerve (Fig. 14.1).

Test and Activation
With the subject lying supine, ask to plantar flex the ankle. Apply pressure against the plantar forefoot (Fig. 15.3a).

Needle Insertion
One handbreadth below the popliteal crease on the medial calf (Fig. 15.3b).
 Caveat: if the needle is inserted too deeply, it will be in the soleus, which is innervated by the same nerve (tibial) and roots (S1, S2) (Fig. 15.3c).

Comments

- The GCM is difficult to fully activate. Activation can be more easily accomplished by flexing the knee first and then asking to plantar flex the ankle.
- The GCM may be involved in lesions of the tibial nerve, sciatic nerve, lumbosacral plexus, and S1 and S2 roots. For the assessment of S1 involvement, the GMH is preferred over the lateral gastrocnemius, which receives some innervation by L5.

15.4 Soleus

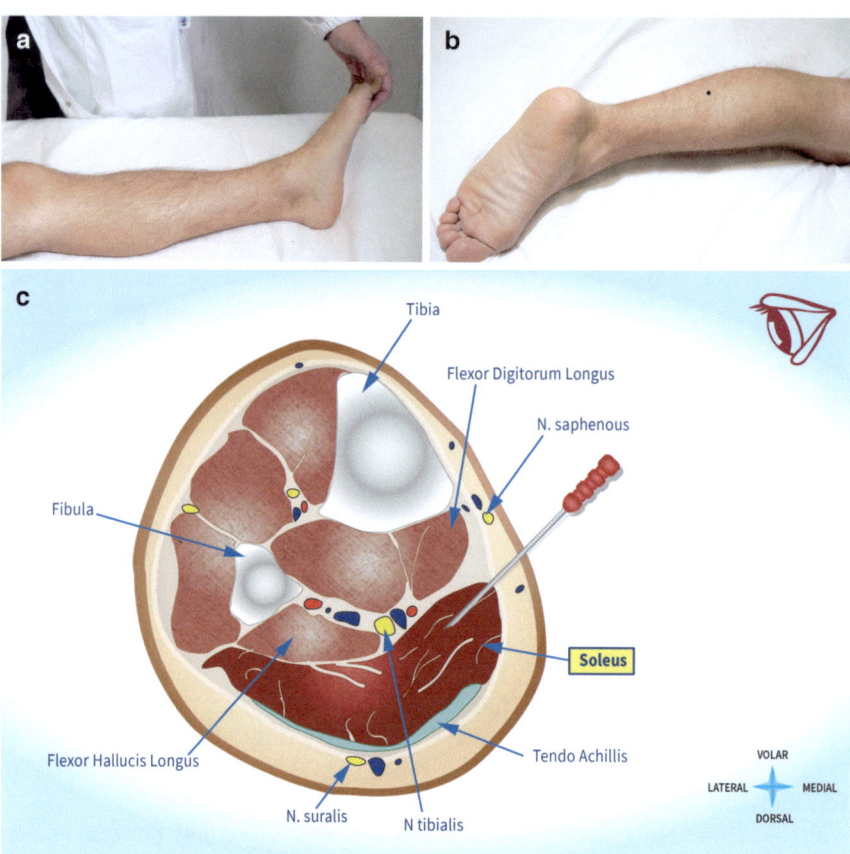

Fig. 15.4 Soleus: (**a**) test, (**b**) insertion point, (**c**) cross-section through the middle of the leg. The eye shows the perspective of the examiner

15.4 Soleus

Innervation and Nerve Fibers Route
S1, S2, lumbosacral plexus, sciatic nerve, tibial nerve (Fig. 14.1).

Test and Activation
With the subject supine, ask to plantar flex the ankle. Apply pressure against the plantar forefoot (Fig. 15.4a).

Needle Insertion
Slightly distal to the midpoint between the knee and the ankle, medially to the Achilles' tendon (Fig. 15.4b).

Caveat: if the needle is inserted too proximally and superficially, it will be in the gastrocnemius; if the needle is inserted too anteriorly pointing toward the tibia, it will be in the flexor digitorum longus (Fig. 15.4c). However, both muscles are innervated by the tibial nerve.

Comments

- The soleus is difficult to fully activate.
- The soleus is a distal S1-innervated muscle.
- The soleus is used to study the H reflex.

15.5 Tibialis Posterior

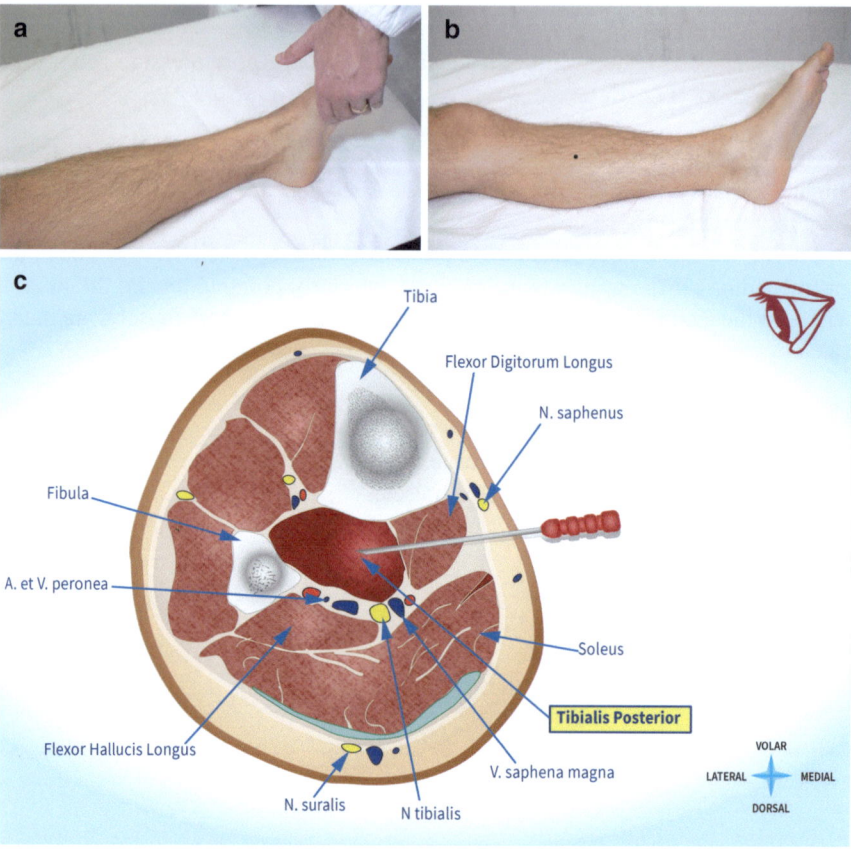

Fig. 15.5 Tibialis posterior: (**a**) test, (**b**) insertion point, (**c**) cross-section through the middle of the leg. The eye shows the perspective of the examiner

Innervation and Nerve Fibers Route
L5, S1, lumbosacral plexus, sciatic nerve, tibial nerve (Fig. 14.1).

Test and Activation
With the subject supine, ask to invert the foot with plantar flexion. Apply pressure against the medial side and plantar surface of the foot in the direction of dorsiflexion and eversion (Fig. 15.5a).

Needle Insertion
One handbreadth distal to the tibial tuberosity and one fingerbreadth off the medial edge of the tibia (Fig. 15.5b). The needle is directed obliquely through the soleus and flexor digitorum longus, just posterior to the tibia.

15.6 Abductor Hallucis Brevis (AHB)

Caveat: if the needle is inserted too superficially, it will be in the soleus or flexor digitorum longus, both innervated by the tibial nerve (Fig. 15.5c). If the needle is pointed too posteriorly, the tibial nerve and the adjacent vessels may be injured.

Comments

- The tibialis posterior is the deepest muscle in the posterior compartment, using a 50 mm long needle for examination.
- The tibialis posterior is a predominantly L5-innervated tibial muscle and is very useful in the evaluation of foot drop to differentiate a peroneal neuropathy (it is normal) from L5 radiculopathy (it is abnormal).

15.6 Abductor Hallucis Brevis (AHB)

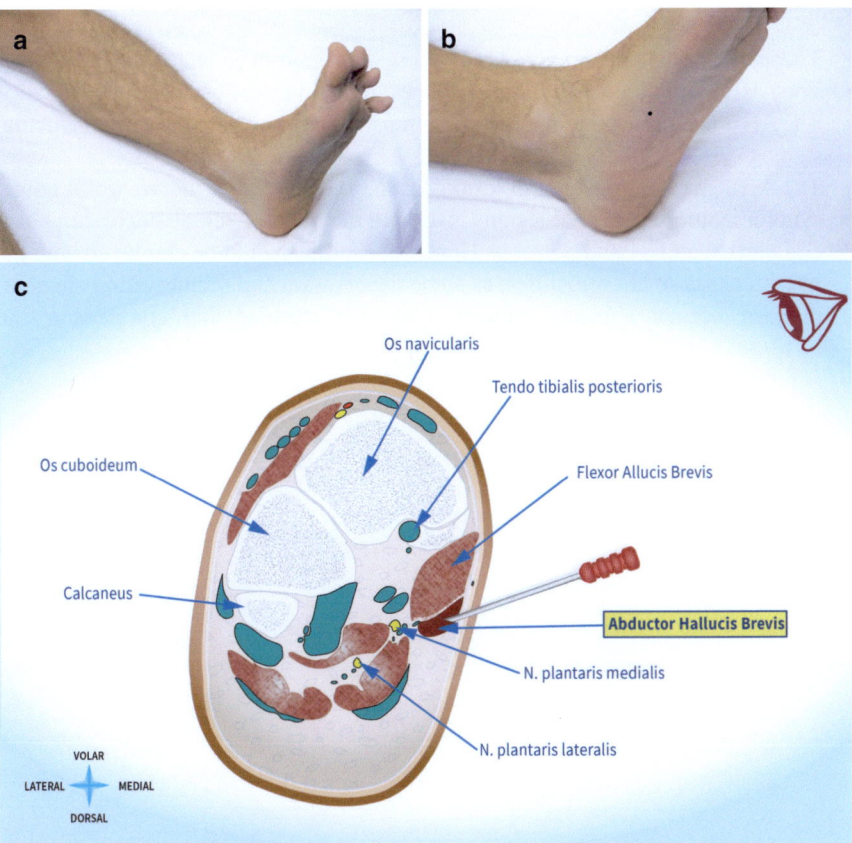

Fig. 15.6 Abductor hallucis brevis: (**a**) test, (**b**) insertion point, (**c**) cross-section through the lower third of the navicular bone, the upper portion of the cuboid and the lower portion of the calcaneum. The eye shows the perspective of the examiner

Innervation and Nerve Fibers Route
S1, S2, lumbosacral plexus, sciatic nerve, tibial nerve, medial plantar nerve (Fig. 14.1).

Test and Activation
With the subject supine, ask to spread their toes (Fig. 15.6a).

Needle Insertion
On the midportion of the medial aspect of the foot between the heel and head of the first metatarsal bone (Fig. 15.6b).

Caveat: if the needle is inserted too distally, it will be in the flexor hallucis brevis; if inserted too deeply, it will be in the flexor digitorum brevis, both innervated by the tibial nerve (Fig. 15.6c). If the needle is inserted too deeply, it may injure the medial plantar nerve.

Comments
- The abductor hallucis brevis (AHB) is often difficult to activate.
- The AHB is often felt painful at needle examination.
- The AHB is employed as the recording muscle for tibial nerve motor conduction studies.
- The AHB may be involved in lesions of the medial plantar nerve, tarsal tunnel syndrome, more proximal lesions involving the tibial or sciatic nerve, lumbosacral plexus, and S1 and S2 roots. However, caution should be employed with the interpretation of abnormalities, as some denervation and reinnervation are common in normal subjects without symptoms. Side-to-side comparison can be useful.

15.7 Abductor Digiti Quinti Pedis (ADQP)

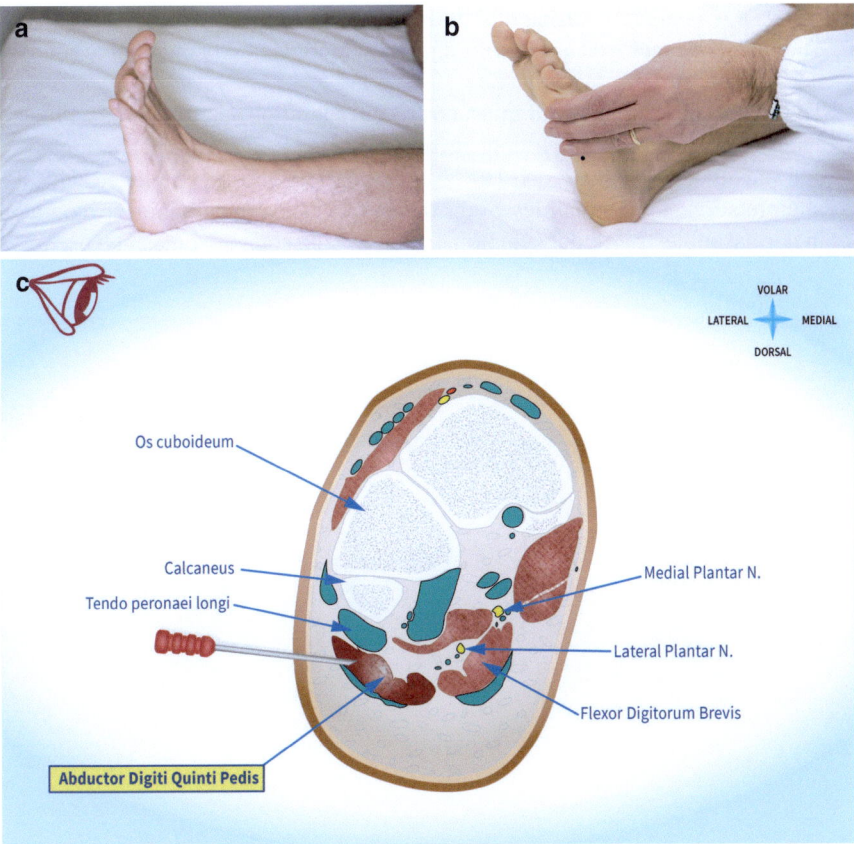

Fig. 15.7 Abductor digiti quinti pedis: (**a**) test, (**b**) insertion point, (**c**) cross-section through the lower third of the navicular bone, the upper portion of the cuboid and the lower portion of the calcaneum. The eye shows the perspective of the examiner

Innervation and Nerve Fibers Route
S1, S2, lumbosacral plexus, sciatic nerve, tibial nerve, lateral plantar nerve (Fig. 14.1).

Test and Activation
With the subject supine, ask to spread their toes (Fig. 15.7a).

Needle Insertion
On the lateral border of the foot, two to three fingerbreadths proximal to the fifth metatarsophalangeal joint (Figs. 15.7b, c).

Comments

- The abductor digiti quinti pedis (ADQP) is often difficult to activate.
- The ADQP is often felt as painful during needle examination.
- The ADQP is employed as the recording muscle for lateral plantar nerve conduction studies.
- The ADQP may be involved in lesions of the lateral plantar nerve, tarsal tunnel syndrome, more proximal lesions involving the tibial or sciatic nerve, lumbosacral plexus, and S1-S2 roots. However, caution should be employed with the interpretation of abnormalities. Some denervation and reinnervation are common in normal subjects without symptoms. Side-to-side comparison can be useful.

Muscles Innervated by the Sciatic and Peroneal Nerves

16

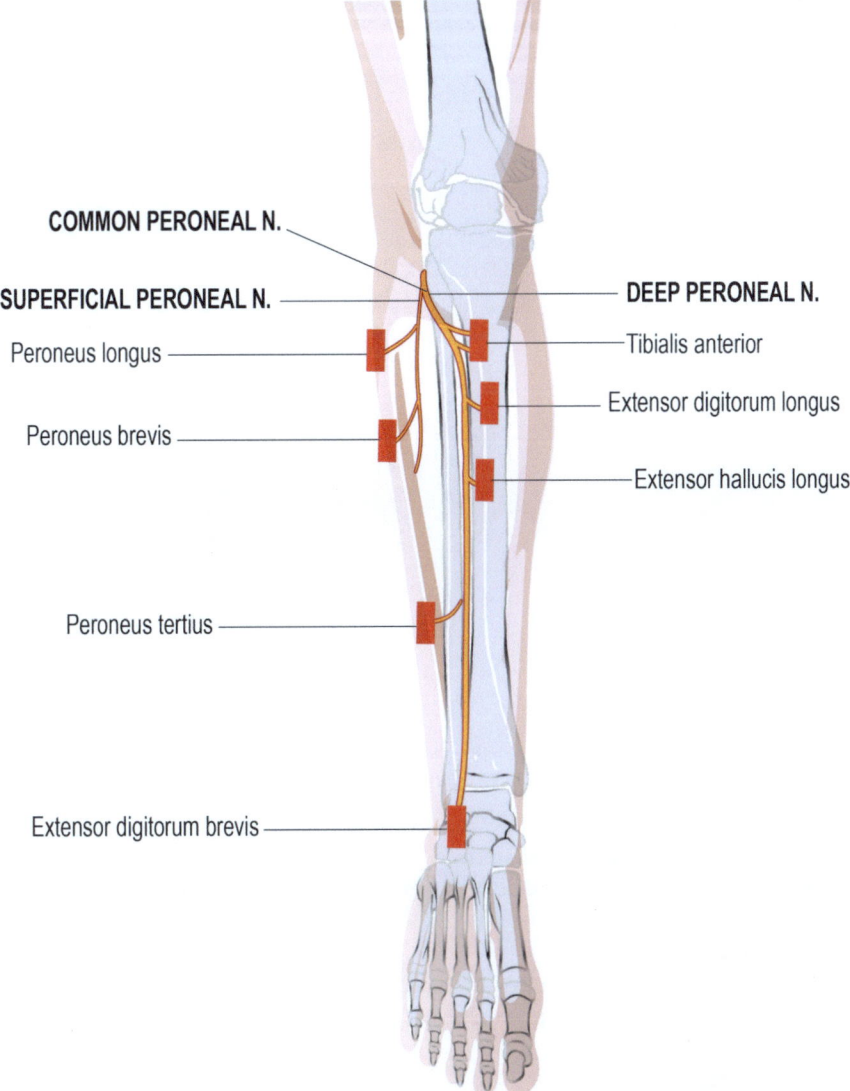

Fig. 16.1 The common peroneal, the superficial and deep peroneal nerves and the muscles they supply. For the innervation of the short head of the biceps femoris by the peroneal division of the sciatic nerve see Fig. 14.1

16.1 Biceps Femoris-Caput Brevis (BF-CB)

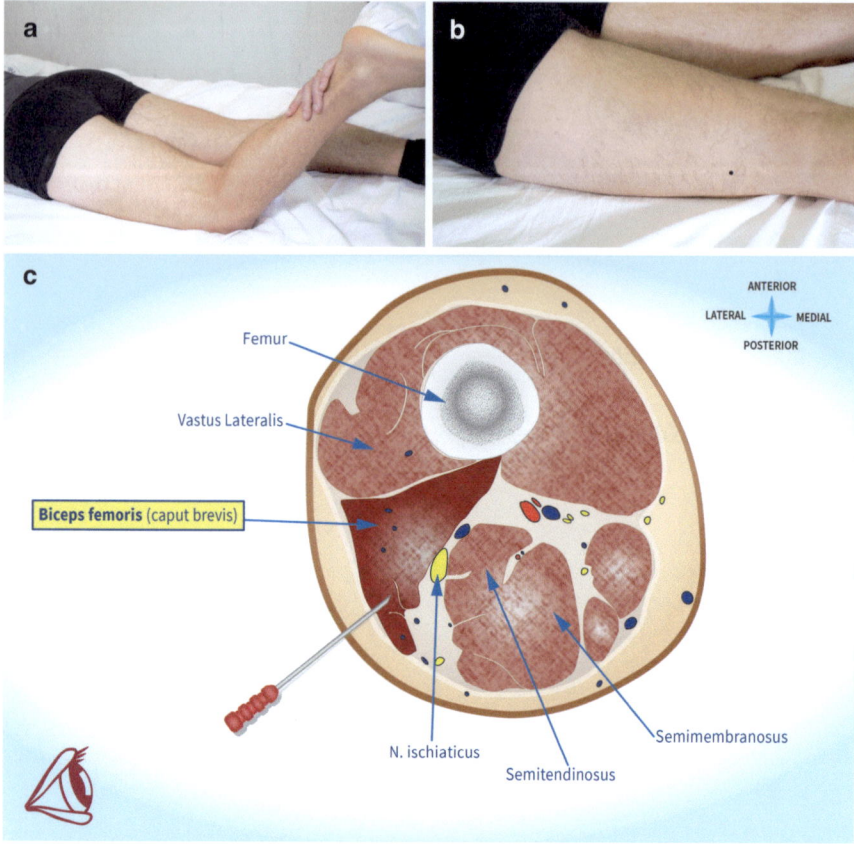

Fig. 16.2 Biceps femoris-caput brevis: (**a**) test, (**b**) insertion point, (**c**) cross-section anatomy through the lower third of the thigh. The eye shows the perspective of the examiner

Innervation and Nerve Fibers Route
L5, S1, lumbosacral plexus, sciatic nerve (peroneal division) (Fig. 14.1).

Test and Activation
With the subject prone, ask to flex the knee. Apply pressure against the leg in the direction of knee extension (Fig. 16.2a).

Needle Insertion
Three to four fingerbreadths proximal to the fibular head and medial to the tendon to the long head of the biceps femoris (Fig. 16.2b).

Caveat: if the electrode is inserted too medially, it will be in the semimembranosus; if inserted too proximally, it will be in the long head of the biceps femoris, both innervated by the tibial nerve (Fig. 16.2c). The muscle is superficial; if the needle is directly inserted too medially and deeply, the sciatic nerve may be injured.

Comment

- The biceps femoris-caput brevis (BF-CB) is the only muscle innervated by the peroneal division of the sciatic nerve above the fibular head. It is important to examine as it will be normal in lesions of the peroneal nerve at the fibular head or lower, whereas it may be abnormal in sciatic nerve lesions that clinically mimic a peroneal neuropathy.

16.2 Peroneus Longus

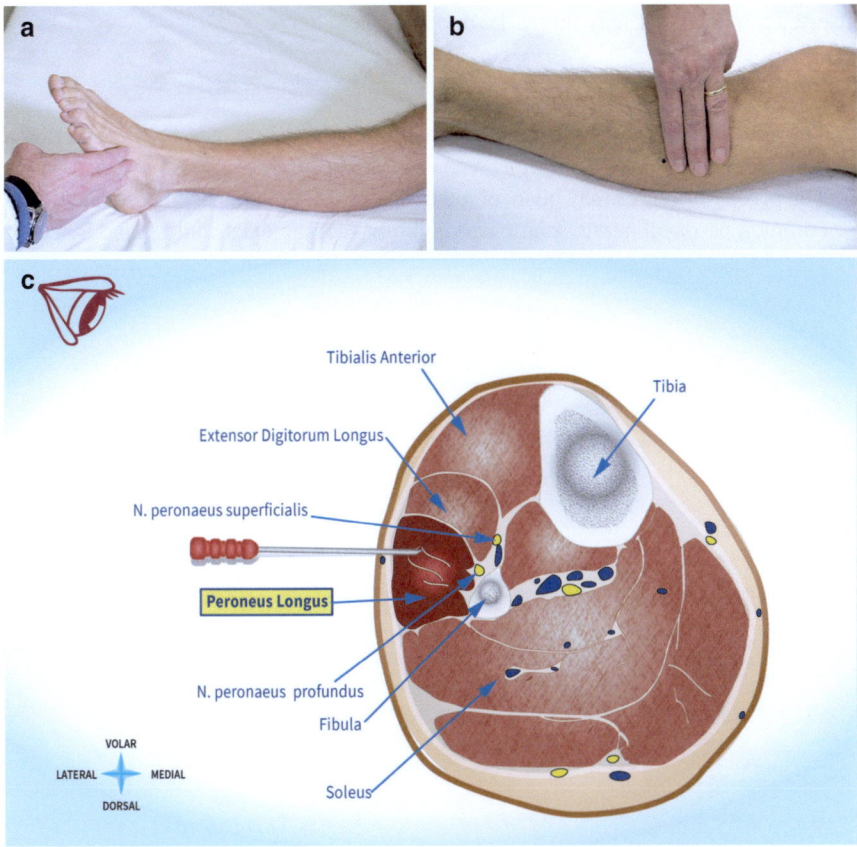

Fig. 16.3 Peroneus longus: (**a**) test, (**b**) insertion point, (**c**) cross-section through the upper third of the leg. The eye shows the perspective of the examiner

Innervation and Nerve Fibers Route

L5, S1, S2, lumbosacral plexus, sciatic nerve, common peroneal nerve, superficial peroneal nerve (Fig. 16.1).

Test and Activation

With the subject lying supine, ask to evert the foot with plantar flexion of the ankle. Apply pressure against the lateral border and sole of the foot in the direction of inversion (Fig. 16.3a).

Needle Insertion

Three fingerbreadths distal to the fibular head (Fig. 16.3b).

Caveat: if the needle is inserted too posteriorly, it will be in the soleus innervated by the tibial nerve; if inserted too anteriorly, it will be in the extensor digitorum longus (EDL) innervated by the deep peroneal nerve (Fig. 16.3c). If the needle is inserted too deeply, it may injure the deep peroneal nerve.

Comments

- The peroneus longus is the most accessible muscle innervated by the superficial peroneal nerve.
- The peroneus longus may be involved in lesions of the superficial peroneal nerve, common peroneal nerve, sciatic nerve, lumbosacral plexus, and L5 and S1 roots.

16.3 Tibialis Anterior

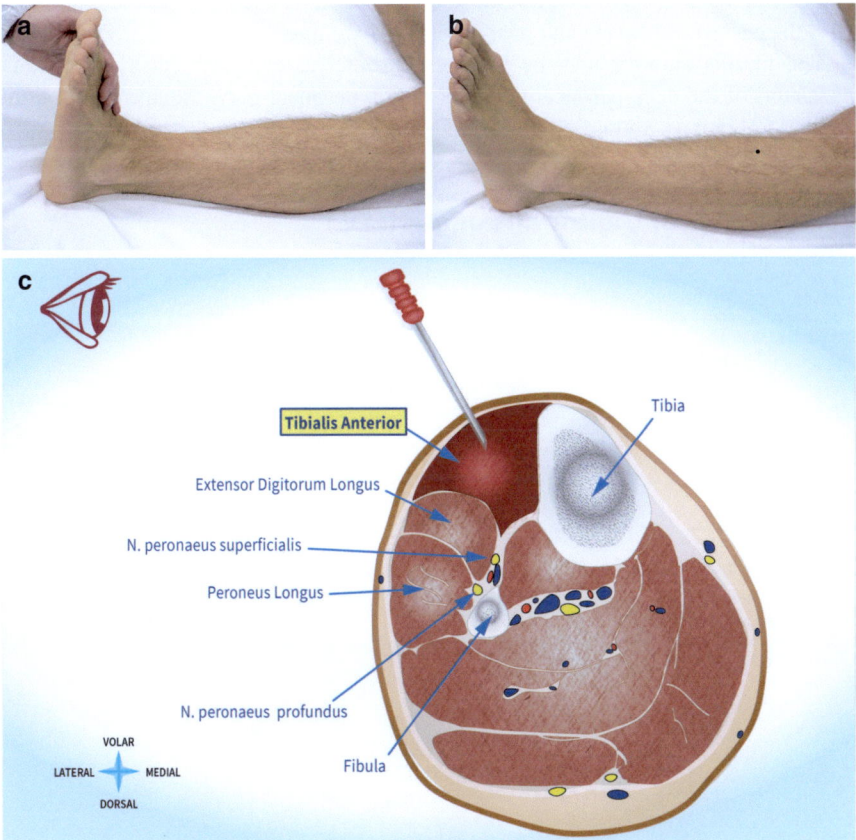

Fig. 16.4 Tibialis anterior: (**a**) test, (**b**) insertion point, (**c**) cross-section through the upper third of the leg. The eye shows the perspective of the examiner

Innervation and Nerve Fibers Route
L4, L5, lumbosacral plexus, sciatic nerve, common peroneal nerve, deep peroneal nerve (Fig. 16.1).

Test and Activation
With the subject lying supine, ask to dorsiflex the foot. Apply pressure against the dorsal and medial surface of the foot in the direction of plantar flexion and eversion (Fig. 16.4a).

Needle Insertion
Four fingerbreadths below the tibial tuberosity and one fingerbreadth lateral to the tibial cresta (Fig. 16.4b).

Caveat: if the electrode is inserted too laterally and deeply, it will be in the extensor digitorum longus, also innervated by the deep peroneal nerve (Fig. 16.4c).

Comments

- The tibialis anterior is the first muscle innervated by the deep peroneal nerve; it is the easiest to activate and the simplest to examine.
- The tibialis anterior may be involved in anterior compartment syndrome, in lesions of the deep peroneal nerve, common peroneal nerve, sciatic nerve, lumbosacral plexus, and L4 and L5 roots.

16.4 Extensor Digitorum Longus (EDL)

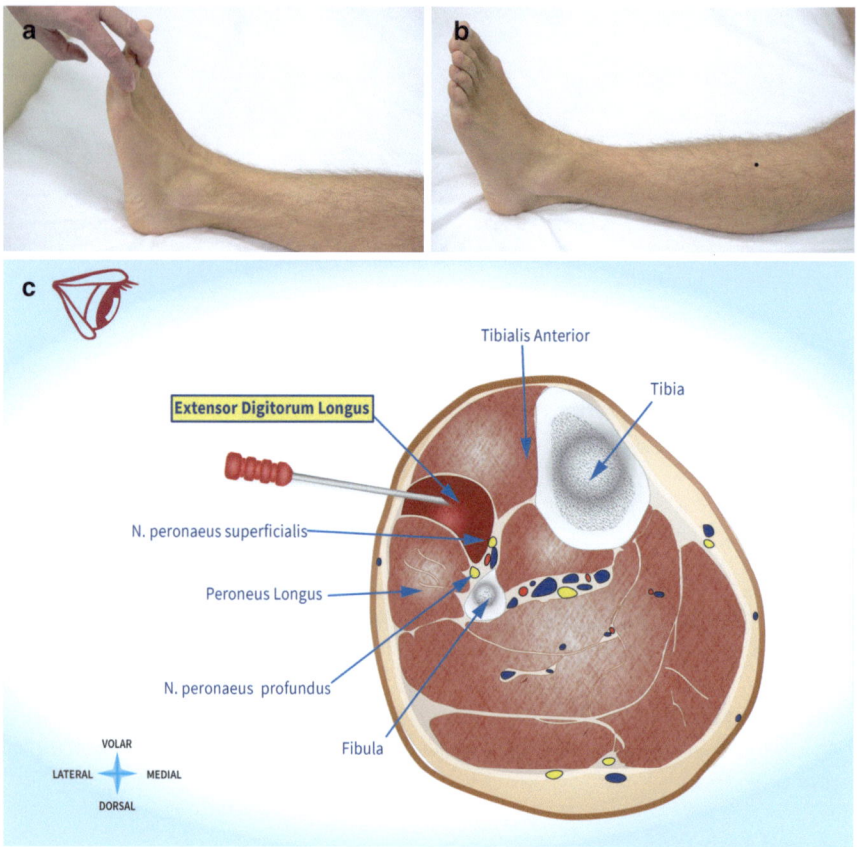

Fig. 16.5 Extensor digitorum longus. (**a**) test, (**b**) insertion point, (**c**) cross-section through the upper third of the leg. The eye shows the perspective of the examiner

Innervation and Nerve Fibers Route
L4, L5, S1, lumbosacral plexus, sciatic nerve, common peroneal nerve, deep peroneal nerve (Fig. 16.1).

Test and Activation
With the subject lying supine, ask to extend their toes. Apply pressure against the dorsal surface of the toes in the direction of flexion (Fig. 16.5a).

16.5 Extensor Hallucis Longus (EHL)

Needle Insertion

Four fingerbreadths distal to the tibial tubercle and two fingerbreadths lateral to the tibial crest (Fig. 16.5b).

Caveat: if the electrode is inserted too medially, it will be in the tibialis anterior; if the needle is inserted too laterally, it will be in the peroneus longus (Fig. 16.5c). Both muscles are innervated by the deep peroneal nerve.

Comment

- The EDL may be abnormal in anterior compartment syndrome, lesions of the deep peroneal nerve, common peroneal nerve, sciatic nerve, lumbosacral plexus, and L5 and S1 roots.

16.5 Extensor Hallucis Longus (EHL)

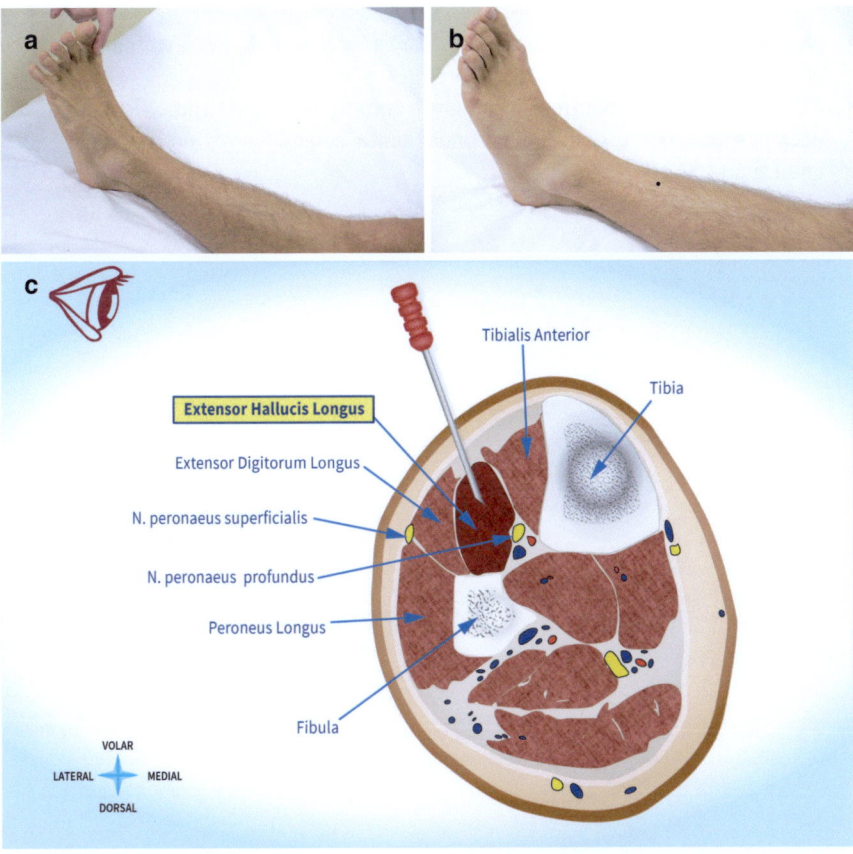

Fig. 16.6 Extensor hallucis longus: (**a**) test, (**b**) insertion point, (**c**) cross-section through the lower third of the leg. The eye shows the perspective of the examiner

Innervation and Nerve Fibers Route
L4, L5, S1, lumbosacral plexus, sciatic nerve, common peroneal nerve, deep peroneal nerve (Fig. 16.1).

Test and Activation
With the subject lying supine, ask to extend their big toe. Apply pressure against the dorsal surface of the big toe in the direction of flexion (Fig. 16.6a).

Needle Insertion
Three fingerbreadths above the bimalleolar line of the ankle just lateral to the crest of the tibia (Fig. 16.6b).

Caveat: if the electrode is inserted too proximally, it will be in the tibialis anterior; if the needle is inserted too laterally, it will be in the peroneus tertius (Fig. 16.6c). Both muscles are innervated by the deep peroneal nerve.

Comments
- Needle examination of the extensor hallucis longus (EHL) is often painful as the needle passes through several tendons.
- The EHL may be abnormal in anterior compartment syndrome, lesions of the deep peroneal nerve, common peroneal nerve, sciatic nerve, lumbosacral plexus, and L5 and S1 roots.

16.6 Extensor Digitorum Brevis (EDB)

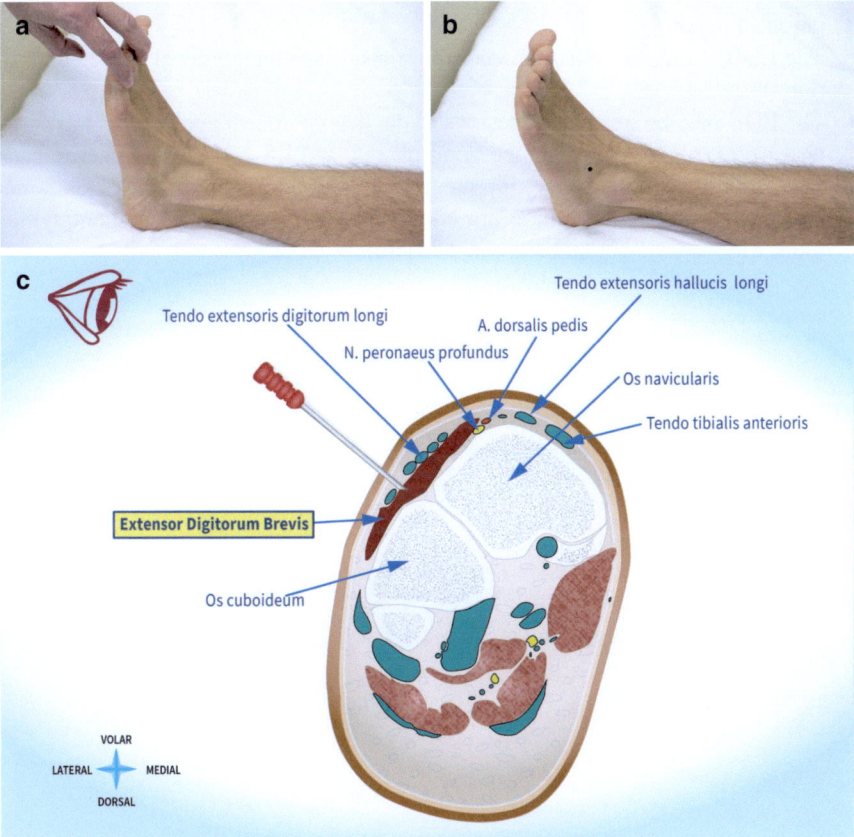

Fig. 16.7 Extensor digitorum brevis: (**a**) test, (**b**) insertion point, (**c**) cross-section through the lower third of the navicular bone, the upper portion of the cuboid and the lower portion of the calcaneum. The eye shows the perspective of the examiner

Innervation and Nerve Fibers Route
L4, L5, S1, lumbosacral plexus, sciatic nerve, common peroneal nerve, deep peroneal nerve (Fig. 16.1).

Test and Activation
With the subject lying supine, ask to extend their toes. Apply pressure against the metatarsophalangeal joints of digits 2–4 (Fig. 16.7a).

Needle Insertion
Three fingerbreadths distal to the lateral malleolus (Fig. 16.7b, c). The muscle is superficial and may be thin. The tendons of extensor digitorum longus run over the muscle.

Comments

- The extensor digitorum brevis (EDB) is the last muscle innervated by the deep peroneal nerve.
- The EDB is employed as the recording muscle for the peroneal nerve conduction study.
- The EDB may be involved in lesions of the deep peroneal nerve, common peroneal nerve, sciatic nerve, lumbosacral plexus, and L5–S1 roots. However, caution should be employed with the interpretation of abnormalities as some denervation and reinnervation are common in normal subjects without symptoms. Side-to-side comparison can be useful.

Paraspinal Muscles

17.1 Cervical

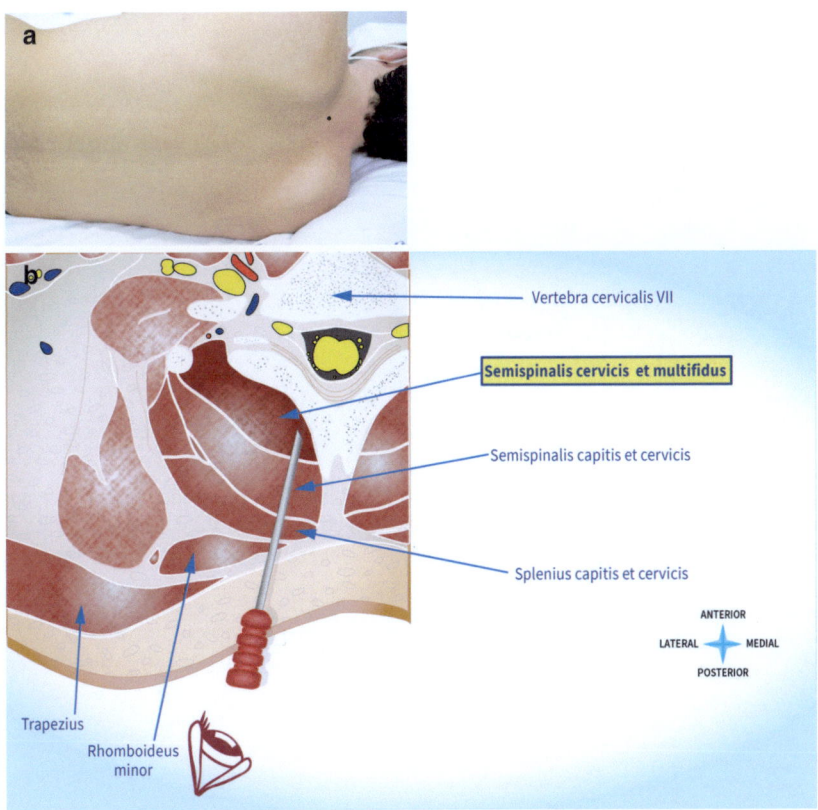

Fig. 17.1 Cervical paraspinal muscles: (**a**) insertion point, (**b**) cross section through the upper margin of the body, the posterior arch, and the transverse process of the seventh cervical vertebra. The eye shows the perspective of the examiner

17.2 Thoracic

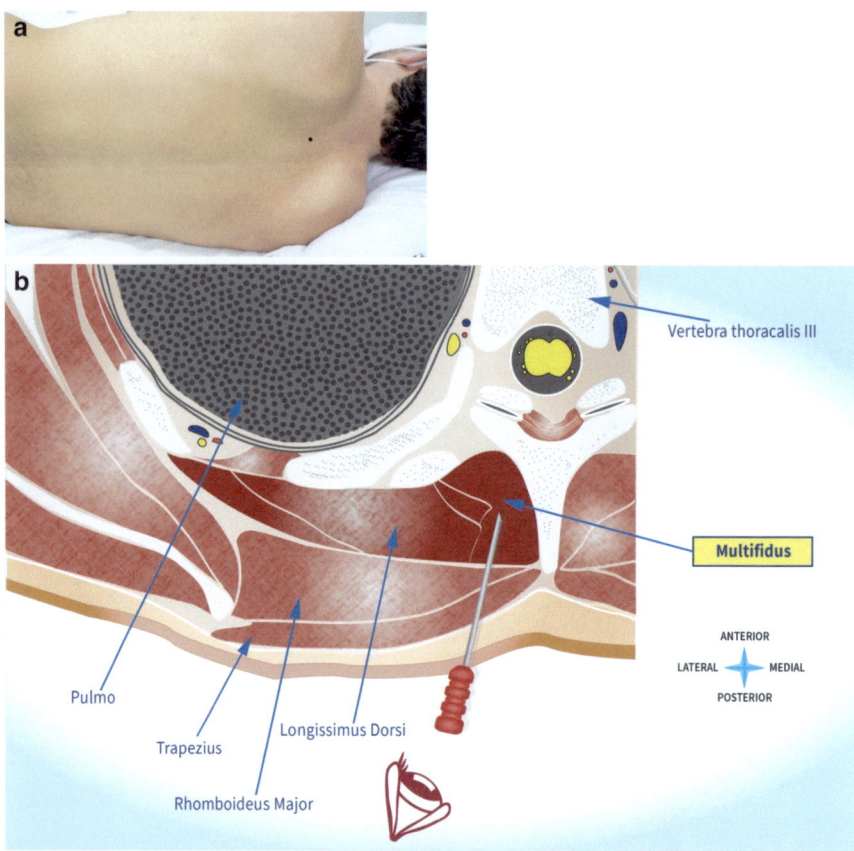

Fig. 17.2 Thoracic paraspinal muscles: (**a**) insertion point, (**b**) cross section through the upper margin of the body, the posterior arch, and the transverse process of the third thoracic vertebra. The eye shows the perspective of the examiner

17.3 Lumbar

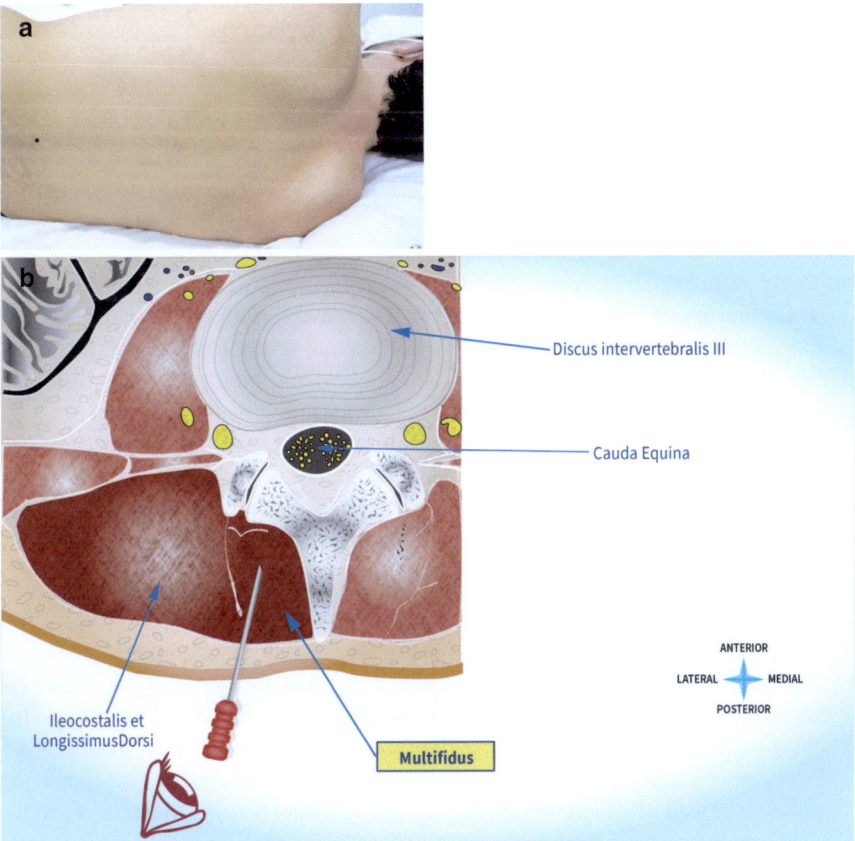

Fig. 17.3 Lumbar paraspinal muscles: (**a**) insertion point, (**b**) cross section through the third lumbar disc and the spinous process of the third lumbar vertebra. The eye shows the perspective of the examiner

Innervation and Nerve Fibers Route
Nerve roots, spinal nerves, dorsal rami. The innervation usually extends one/two segments above and below a segmental level. This generates a significant amount of overlapping in the innervation of paraspinal muscles, especially in the superficial layers.

Test and Activation
The subject is in a prone position. For the cervical paraspinals, ask the subject to elevate the head; for the thoracic paraspinals, ask the subject to extend the back; for the lumbosacral paraspinals, ask the subject to elevate the whole leg (straight from the hip) of the side under study.

Needle Insertion
The subject is lying on their side or in a prone position. Before inserting the electrode, the level of the spine must be identified. For the cervical and thoracic paraspinals, the spinous process of C7 is identified, and the count is done up or down accordingly. For the lumbosacral paraspinals, an imaginary line is drawn between the upper part of the iliac crests. This line intersects the spinal column at the L3–L4 intervertebral level. The count proceeds up or down accordingly.

Insert the needle one finger breadth laterally to the spinous process of the identified level until reaching the lamina of the vertebra, and then pull back slightly to ensure testing the multifidus muscle that has the least innervation overlap.

Caveat: if the electrode is inserted too superficially, it may be in the superficial muscular layer of the back (trapezius, latissimus dorsi, rhomboids, or splenius). Rare cases of pneumothorax have been reported in assessing the lower cervical and thoracic paraspinal muscles with the needle insertion too lateral.

Comments

- The paraspinal muscles are the most proximal muscles to examine.
- The paraspinal muscles often are difficult to relax for assessing spontaneous activity. To obtain relaxation in cervical paraspinals, a pillow is placed across the chest of the patient, allowing the patient's head to flex and rest on the forehead. For the thoracic area, the best position is flat. For the lumbosacral paraspinal, the pillow is placed across the abdomen, producing a mild bowing of the lower spine.
- Abnormalities in the paraspinal muscles localize the lesion at the root or motor neuron level.
- Paraspinal muscles are segmentally affected in radiculopathies. However, because of the wide overlap of myotomes, the specific root level is best determined by examining the limb muscles.
- The paraspinal muscles can be affected in processes involving the roots and in vascular, inflammatory, and degenerative (amyotrophic lateral sclerosis) conditions involving the anterior horn cells.
- Thoracic paraspinals are important to test in the differential diagnosis between amyotrophic lateral sclerosis (may show spontaneous activity) and cervical and lumbar stenosis in tandem (usually are not affected).

MIX
Papier aus verantwortungsvollen Quellen
Paper from responsible sources
FSC® C105338

If you have any concerns about our products,
you can contact us on
ProductSafety@springernature.com

In case Publisher is established outside the EU,
the EU authorized representative is:
Springer Nature Customer Service Center GmbH
Europaplatz 3, 69115 Heidelberg, Germany

Printed by Libri Plureos GmbH
in Hamburg, Germany